RELATIVITY VISUALIZED

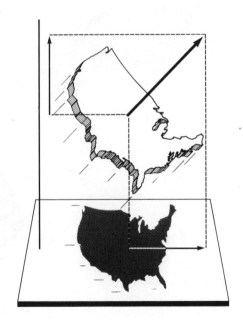

Insight Press

614 Vermont Street, San Francisco, CA 94107

RELATIVITY VISUALIZED

Written and Illustrated by

Lewis Carroll Epstein

City College of San Francisco

INSIGHT PRESS · SAN FRANCISCO

Library of Congress catalog card #82-084280

ISBN 0-935218-05-X

Insight Press
614 Vermont Street
San Francisco, CA 94107

In Memory of My Father

HARRY EPSTEIN
Attorney at Law
3229 MISSION ST., AT VALENCIA
SAN FRANCISCO

The following people played important parts in putting this story together:

encouragement	**Michael Scott** **Robert Pisani**
education	**Michael Gardebled** **Jonathan Kern** **Louise Ainsworth** **Gerald Leone**
discussions of theory	**Harry Dart** **John Ehlers** **David Wall** **Robert Nason** **Peter Epstein** **Danny Epstein** **Renie Baus** **David Sonnabend**
word processing	**Margaret Oakley**
organizing	**Amy McManus**
initiation to the world of publication	**Robert Head** **Darlien Fife** **Paul Hewitt**
illustrations	**Ellen Buoncristiani**
editing, design, and production	**Robyn Brode**
typesetting	**Turnaround**
printing	**Fairfield Graphics**

Chapter 5 is the centerpiece of this work. The chapters preceding it relate to its development and those following it relate to application of its perspective.

Table of Contents

Chapter 3
Pandora's Box
Mostly about time desynchronization and its consequences

Chapter 4
Measuring the Consequences
Mostly about how to calculate the effect exactly

Chapter 5
The Myth
Mostly about why space shrinks, time slows, and
why you can't go faster than light

Chapter 6
The Big Bang

Mostly about the fireball radiation, the edge of the universe, and the cosmological principle as applied to the Big Bang cosmology

Chapter 7
The Third Leg

Mostly about how speed affects mass

Chapter 8
$E = mc^2$

Mostly about the weight of energy and what mass is made of

Chapter 12
The Ends of Space and Time
Mostly about the rupturing and mending of heaven

Appendices
Mostly about technical business

While most will sleep, a few will creep, in places deep, where dark things keep...

Chapter 1

The Principle of Relativity

Galileo's Dictum

The year Columbus spent discovering the New World, a university sophomore named Copernicus spent discovering mathematics and astronomy. These discoveries and a lifetime of study led Copernicus to the realization that the earth was not only a globe in space, but that the globe MOVED around the sun.

However, most thinking people found it impossible to believe that the heavy earth could be in motion, indeed, in perpetual motion. They presented reasoned arguments against Copernicus's theory. Here is one such argument: Suppose the earth does move and suppose you drop a coin directly above your toe. By the time the coin falls to earth the earth will have moved and your foot will be carried with it. So the coin must miss your toe. But if you drop a coin above your toe it will, in fact, hit your toe. Thus the earth cannot be moving.

About a century later, while the

Figure 1–1. If the earth was moving, then dropped objects would not fall straight down.

first English colonies were being established in North America, Galileo came to the defense of the moving earth idea. Here, in a nutshell, is Galileo's counterargument. Suppose you are flying along smoothly and standing inside an aircraft cabin, even in a supersonic aircraft, and you drop a coin directly above your toe. Sure enough, it hits your toe.* Galileo referred to a ship's cabin rather than an aircraft cabin, but the central idea is that if you are closed inside a box that is moving smoothly, that is, not starting or stopping or jerking or turning but just moving at one speed in one direction, you **cannot** tell if you are in motion; **everything** in your box happens as if you were at rest. That fundamental idea became known as Galileo's Dictum.

Moreover, if you open a window in the box and look out and see another box approaching your box, you still can't tell if you are moving. You can't tell if you are moving toward the other box or if the other box is moving toward you. All you know is that there is **relative** motion between the boxes.

In essence, Galileo's Dictum is this: All smooth, linear motions are

Figure 1–2. But dropped objects do fall straight down. Therefore, the earth cannot be in motion.

---

*The coin hits your toe because if your toe is moving to the right due to the aircraft's motion, so also is the coin moving to the right at the moment it is dropped, since the coin is joined to the aircraft by your body just before you drop it. The coin keeps moving to the right after you let go. So the coin follows your toe and hits it.

relative. The universe is full of particles, all moving in different directions at different speeds. Who is to say which particle is really, absolutely, at rest? Thus you can never say, "The speed of this particle is 900 miles per hour." All you can say is, "The speed of this particle is 900 miles per hour relative to the air, or relative to the moon, or relative to this earthquake wave front." This is all very reasonable and the business would be permanently settled here were it not for certain curiosities about light first noted by a keen-eyed painter, Franciscus Grimaldi, who was officially employed by the very two popes with whom Galileo had his difficulties.

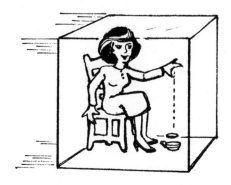

Figure 1–3. Smooth motion in a straight line does not affect what is happening inside the box. You might think that is obvious, yet in the day of Galileo it was a new idea.

Light

Grimaldi observed that surrounding and even inside the shadow of an opaque object were several rings or fringes. Now, most people had supposed that light was a rain of minute particles. That idea fit well into Galileo's scheme of a world of particles and it explained why light flies in straight lines and why shadows are exact silhouettes of the objects that cast them. But it did not explain Grimaldi's fringes.

The artist proposed that light was a fluidlike substance that could flow slightly around and behind objects, and that the fringes resulted from ripples spreading downstream from the edges of an object, just as ripples spread from the edges of a stone

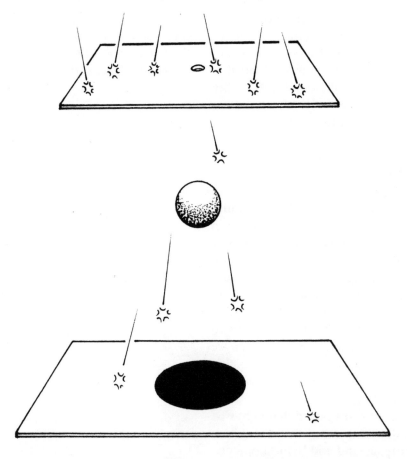

Figure 1–4. If light is a rain of little pellets, an object should completely shield the area behind it, as illustrated here. To do this experiment, you need a point source of light. The sun (or a light bulb) is not a point source. So the light is passed through a pinhole; the pinhole is the point.

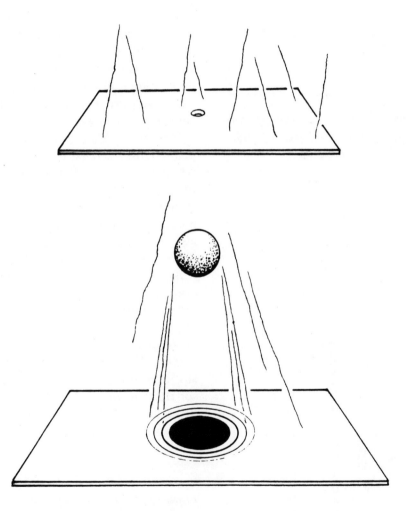

Figure 1–5. But in fact, the area behind an object is not completely shielded. If you look very closely, you will see it is contaminated with rings of light.

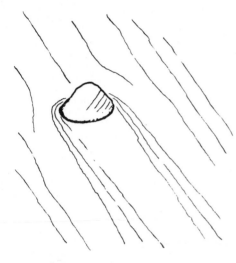

Figure 1-6. Note the waves spreading into the shielded area of quiet water behind the stone.

around which a fast current is flowing. In this way the wave nature of light first came to be suspected.

As the theory was subsequently developed, light was taken to be a vibration transmitted through space, rather than a fluid flow through space. (Yet, flow effects will return us to the mainstream of this story.) The great utility of the wave theory of light was that it showed why light can go around small things—waves spread—and why light sometimes does not appear where you expect it should—waves can cancel each other out.

The wave theory of light explained the halo patterns you see when you look at small, bright lights through a tiny pinhole in a piece of aluminum foil or through a transparent curtain. The wave theory also explained why the steam very close to the stack of a locomotive or teakettle spout is invisible. But most convincing of all, the wave theory led a physics teacher, who was trying to illustrate the wave theory to his class, to build an electric device that would make big demonstration-size light waves. The device Professor Hertz made for the class of 1887 turned out to be the first radio wave transmitter.

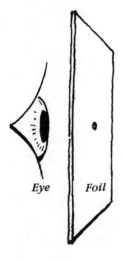

Eye   Foil

Figure 1-7. Make a very tiny pinhole in some aluminum foil. Then go for a night walk and look at very distant streetlights through the hole.

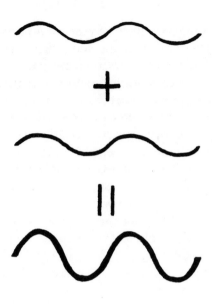

Figure 1–8. These waves are combining so as to reinforce each other. That is, they are added together in phase.

Figure 1–9. These waves are combined so as to cancel each other. That is, they are added together out of phase. The waves are exactly one-half cycle (180 degrees) out of phase, which causes complete destructive interference. But this seems like a violation of conservation of energy. If light cancels light, where does the energy go? It turns out that every time light cancels light at one location there is another location—usually very nearby—where light reinforces light, and all the energy that is missing from the canceled location shows up at the reinforced location. This is true for sound, water, or any other kind of wave.

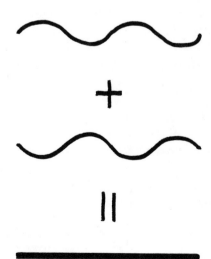

In the Day of the Æther

You can't have water waves without water, you can't have a flag wave without a flag, you can't have sound waves in a vacuum, you can't have earthquake waves without earth, and you can't have an arm wave without an arm. Before the turn of the century, if you were going to take the wave theory of light seriously, and you believed sunlight came from the sun, then you had to believe there must be something in space—in the empty space between the earth and the sun—that could wiggle and thus transmit the light from the sun through the intervening empty space to the earth. That something was called the æther. The name was chosen to depict something on the edge of reality.

If light waves were æther waves, then so also must X rays and radio waves be æther waves. If this was so, then the æther must pass through material objects as well as fill empty space.

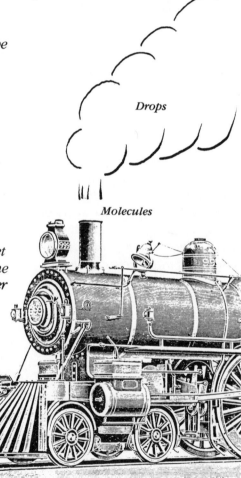

Figure 1–10. Every molecule of water in the steam cloud behind the stack came out of the stack. Yet the water is invisible as it comes out because it comes out in single molecules, which are separated in space. So the light each molecule scatters toward your eye arrives in a different phase, and the different phases cancel each other. But the molecules are sticky, so they stick together to form groups of many thousands of molecules, called droplets. The molecules in a droplet are concentrated in a small volume of space and so the light they scatter has approximately the same phase, resulting in reinforcement, and this makes clouds visible. Phase is a buzz word for describing position along a wave; if the crest of a wave is called zero degrees, the trough is called 180 degrees and the next crest is called 360 degrees.

The æther was a most wonderful substance by means of which not only light, but also gravity and even matter itself, could be accounted for in a **unified** way. Gravity was due to pressure differences in the æther. The pressure in and near massive bodies, like the earth, was supposed to be lower than in distant space. Thus, the æther pressure on the side of the moon facing the earth was less than the pressure on the far side. So there was a net force pushing the moon toward the earth. Likewise, the moon produced a force on the earth.

Atoms were minute vortex whirlwinds (like smoke rings) in the æther. They could spin and thereby hold together forever because the æther was frictionless, like a superfluid.* But if the atom was ever broken, out would spill all the (spin) energy. So the old nineteenth-century theory anticipated that energy would be released if atoms could be broken! The æther was not just some philosophical nonsense idea. It was a working model by which many optical phenomena, such as polarization, diffraction, refraction, and double refraction, could be quantitatively explained.

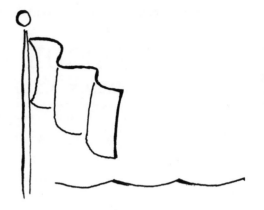

Figure 1–11. A wave is a wiggle and it can exist only if there is something to wiggle. A wave cannot carry the thing that wiggles with it; that is, it cannot be pictured as a kinky noodle flying through space, because waves (like sound, light, or water waves) can pass through each other but noodles can't. The stuff that wiggles must already be in space before the wave arrives.

---

*The vortex holds together because there is lower pressure inside the vortex than outside. There is lower pressure inside because the centrifugal force of spinning resists the pressure of the outside fluid.

Æther atoms as described in the text my father read. It is a wonderful old physics book.

TEXT-BOOK OF PHYSICS

W. WATSON, A.R.C.S., D.Sc. (London), F.R.S.

ASSISTANT PROFESSOR OF PHYSICS AT THE ROYAL
COLLEGE OF SCIENCE, LONDON

LONGMANS, GREEN, AND CO.

91 & 93 FIFTH AVENUE, NEW YORK

LONDON, BOMBAY, AND CALCUTTA

1911

One of the most recent theories, and one which very powerfully appeals to the imagination, is Lord Kelvin's vortex atom theory. By vortex motion is meant a form of motion such as occurs in a smoke-ring. The path of the particles of air in such a smoke-ring is indicated by the arrows in Fig. 95, where the curved arrows show the direction in which the air particles, which are simply rendered visible by the smoke, rotate, while the straight arrow shows the direction in which the ring, as a whole, moves. There is a very important difference between this form of motion and a wave motion. In the latter, although the waves travel onwards, the individual particles of the medium in which the wave is being propagated only move through a comparatively small distance from their original position, the motion being handed on from one particle to the next. In vortex motion, however, the particles of the medium themselves move forward, so that in a smoke-ring the particles of air originally forming the ring move on with the ring.

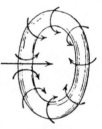

FIG. 95.

The properties of vortex motion were first examined by rigid mathematical methods by von Helmholtz, who found that if the fluid in which this form of motion exists is frictionless, incompressible, and homogeneous, then : (1) A vortex can never be produced, nor if one exists can it be destroyed, so that the number of vortices existing is fixed. This corresponds to the property of indestructibility of matter. (2) The rotating portions of the fluid forming the vortex maintain their identity, and are permanently differentiated from the non-rotating portions of the fluid. (3) These vortex motions must consist of an endless filament in which the fluid is everywhere rotating at right angles to the axis of the filament, unless the filament stretches to the bounding surface of the fluid. (4) A vortex behaves as a perfectly elastic substance. (5) Two vortices cannot intersect each other, neither can a vortex intersect itself.

On the basis of these results of von Helmholtz, Lord Kelvin has founded a theory as to the constitution of matter. He supposes that all space is filled with a frictionless, incompressible, and homogeneous fluid (the ether), and that an atom is simply a vortex in this medium. The existence of different kinds of atoms may be accounted for by the fact that a vortex need not necessarily be a simple ring, as shown in Fig. 95, but might have such a form as that shown in Fig. 96. Since a vortex can never intersect itself, it follows that the number of times such a vortex is linked with itself must always remain the same. Hence we may suppose that the atoms of the different elements are distinguished from one another by the number of times they are linked.

FIG. 96.

The Dictum in Jeopardy

The æther introduced something new into the world of physical ideas. Before æther the world was filled with particles of matter and light, moving with different speeds in every direction. None of these particles had any special status. There was no way to pick some special particle, or group of particles, and say it or they were at rest. But then, along came the æther. It was something special. **It was the underlying substance from which the particles of matter and light were formed.** When you moved through the æther you really moved.

Moreover, the æther was not some distant reference point. It filled all of space. So as the earth moved through space, it seemed reasonable to expect to detect some effect of the æther rushing by. Even in a sealed box you might detect some effect, since the æther passed through material objects, right between the atoms, which were themselves nothing but æther. But if you could detect your motion through space while sealed in a box, then Galileo's Dictum would be smashed.

If the flow of æther could be detected, and it seemed it could, then Galileo's Dictum would be dead. When the possibility of doing such a significant experiment became clear in the last years of the Victorian era, many scientists tried to do it. They tried all kinds of methods. Some of the methods, if successful, would have extracted useful energy from the earth's motion through the æther. However, the conceptually simplest method was to flash a light bulb hanging in the middle of a room and time how long the light took to get from the bulb to the room's walls. If the room was moving through the æther, then the expanding æther wave (light wave) created by the flash should be left behind as the room moved forward, and so the light should reach the room's trailing wall before it reached the room's leading wall. But when that kind of experiment was tried, the light always reached the two opposite walls simultaneously, **as timed by experiments in the room,** regardless of the room's motion. The expected æther flow effect was surprisingly **not** detected.*

The Dictum Prevails

Not only this kind of experiment, but also all the methods of detecting the æther flow failed when tested by trial in laboratories. No one knew why all the different æther flow experiments failed. There was no theory to explain what was happening.

---

*Because of purely engineering problems the conceptually simplest experiment is technically the most difficult to do. However, introductory books on relativity and general physics describe in detail how technically simpler experiments were conducted. I suggest you make the time to read about the actual experiments.

Figure 1–12. Motion picture films are read from bottom to top; the films start at the bottom and finish at the top. These films show a light bulb hanging in the middle of a room being switched on. The room has a glass side so that you, outside, can see what is going on inside the room. The first film shows the light wave expanding from the flash when the room is not moving. As expected, the flash reaches opposite walls simultaneously. The second film shows the bulb being switched on when the room is in motion through the æther. If the room moves to the right through the æther, then the æther must seem to flow to the left through the room. The part of the flash moving to the left can take a free ride on this flow. The flow acts as a tailwind and enables the light to make it to the left wall in one frame. The part of the flash moving to the right, against the direction the æther is flowing through the room, requires three

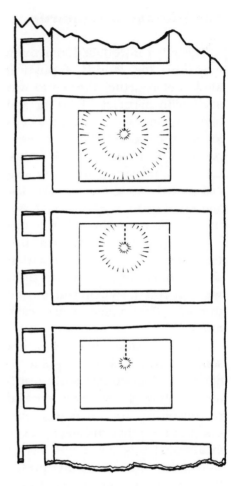

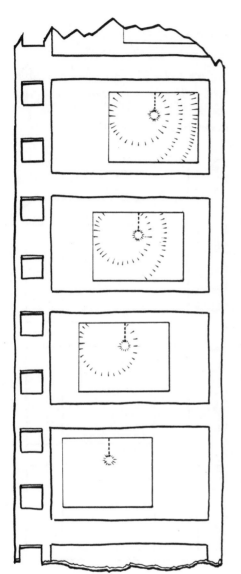

frames to make it to the right wall.
BUT WHEN EXPERIMENTERS **IN**
THE ROOM TRY THIS EXPERIMENT,
THEY FIND THAT THE LIGHT
FLASH ALWAYS GETS TO THE LEFT
AND RIGHT WALLS IN EXACTLY
THE SAME TIME, REGARDLESS OF
THE ROOM'S MOTION. *That is,
experimenters riding within the
room always find the sequence of
events to be as illustrated in the first
film. But if you were outside and
watched the happening inside as the
room* **flew past** *you, would you
find the sequence of events illustrated in the first film or the sequence
illustrated in the second film? Is it
conceivable that experimenters
inside the room would find the
sequence of events illustrated in the
first film, while those watching the
same happening from outside the
moving room would find the
sequence of events illustrated in the
second film? If that is conceivable,
the world must have some unexpected kinks in it, but perhaps it
does.*

But the laboratory is the supreme court of physics where God renders judgment, and in the laboratory no experiment could defeat Galileo's Dictum.

Gradually, it began to occur to some insightful people that basic theory would have to be altered to be compatible with a new fundamental principle. This new principle explicitly excluded the possibility that anything, even light, could ever be expected to circumvent Galileo's Dictum. In 1904, during a talk in Saint Louis, Missouri, Henri Poincaré, one of the most respected scientists of the age and cousin to the president of France, coined the term the Principle of Relativity for the proposed new fundamental principle. But he did not say how to alter physics in order to make it compatible with the new principle.

The understanding of this situation did not improve until an acute crisis came to be appreciated. The crisis was caused by yet another, far more obvious and embarrassing problem, the central problem, which will be the subject of the next chapter.

Summary

According to Galileo's Dictum, a coin dropped directly above your toe will hit your toe, even if you are in a moving room. Likewise, you might expect that when a room light is switched on the flash will first hit the floor directly under the

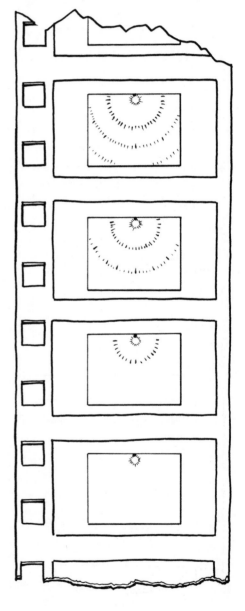

Figure 1-13. This film shows a light bulb being turned on in a stationary room. The light flash first hits the floor directly under the bulb.

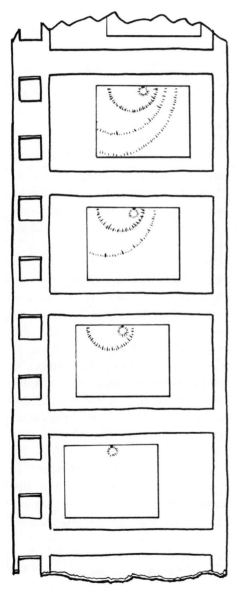

light bulb, even if you are in a moving room. However, if light is æther waves and the room is moving through the æther, which means that an æther wind is blowing through the room, the light should be deflected by the wind and therefore should **not** first hit the floor directly under the bulb, or so most scientists thought. When experimenters put this idea to the test **inside** the room, they found that the light, like the coin, was in fact not deflected and did indeed first hit the floor directly under the bulb, regardless of the room's motion and regardless of what most scientists thought.

The extension of Galileo's Dictum to include aetherial things (like light), as well as mechanical things (like coins), was called the Principle of Relativity.

*Figure 1–14. This film shows a light bulb being turned on in a room that is moving by you. It shows the light deflected by the æther wind blowing through the room, and so the flash does not first hit the floor directly under the bulb. BUT WHEN EXPERIMENTERS IN THE ROOM TRY THIS, THEY, IN FACT, FIND THAT THE FLASH DOES FIRST HIT THE FLOOR DIRECTLY UNDER THE BULB, REGARDLESS OF MOTION. If the room had a glass wall, and you watched what was going on inside as the room **flew past** you, would you also see the flash first reach the floor directly under the bulb? Perhaps not.*

Figure 1–15.

Question

Three spacecraft are traveling in single-file formation, with equal separation between ships. They are in deep space far from any nearby reference. The astronauts desire to know their speed. To estimate their speed the middle ship sends out a radio signal or light flash. If the ships are not moving, the signal will reach the lead ship and the rear ship simultaneously. If the formation is in motion, can its speed be inferred from how much more time it takes the signal to reach (by chasing) the lead ship than to reach (by running into) the rear ship?

a Yes, speed can be inferred as described above.

b No, speed cannot be inferred as described above.

Answer

The answer is **b**. Suppose for a moment the method would work. Then you could surround the fleet with a big box, moving at the same speed the ships are moving, and inside that box the astronauts could tell that they, their ships, and the box were moving through space. But if they could tell they were moving it would be a direct violation of the Principle of Relativity. So the astronauts must find that the signal always takes as much time to reach the lead ship as it takes to reach the rear ship. Note that this is what the astronauts riding the spacecraft must find. It does not dictate what you watching the spacecraft fly by might find. But you must admit, it would be a strange world if the astronauts found that the signals arrived at the lead and rear ship simultaneously and you found they did not arrive simultaneously. Can you trust your intuition on this?

The old æther. What sort of substance was the æther supposed to be? First think about air and sound. Sound is a compression wave in the air; layers of air are alternately compressed and decompressed. The compression wave gets started because something moves back and forth. Sound comes, for example, out of the end of a horn and travels in the direction of the compression.

Æther waves also get started because something moves back and forth. For example, radio waves (which are nothing but long light waves) are started because electrons move back and forth in the transmitter antenna. But the radio wave does not come out of the end of the antenna. It comes out of the side of the antenna. So the radio wave, unlike the sound wave, travels in a direction at right angles to the back and forth motion. In the illustration, the layers of æther alternately move up and down as the wave runs away from the antenna.

Figure 1–16.

The wave runs because each layer acts on its neighbor. Each layer exerts a shearing force on the next layer. If one layer is to exert a shearing force on the next layer there must be a solid connection between them. This meant that the æther had to be pictured as a rigid elastic substance. (Steel, for example, is both rigid and elastic.) Why could the æther not be pictured as a gas or liquid? Because gas and liquid have no shear strength. That is why you can poke your finger into water but not into steel. Shear waves cannot travel in a substance that has no shear strength. (The center of the earth is known to be liquid because shear-type earthquake waves cannot travel through it.) Yet the æther could not be rigid. Why? Because the planets move through it.

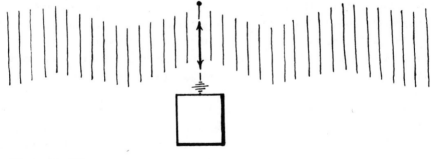

Figure 1–17.

The æther had to be a rigid elastic fluid. You might think this is an impossible combination. But, as Lord Kelvin emphasized, it was a possible combination. When deformed slowly a rigid fluid acts as a fluid. When rapid deformation is attempted it becomes rigid. So for the rapidly moving light waves the æther could be rigid, and for the planets that move slowly, when compared to light, it could be fluid. Æther was not the only rigid fluid in the world.

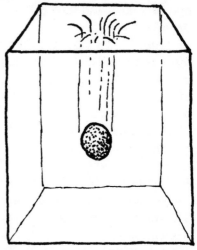

Ice is a rigid fluid, since a stone put on top of a block of ice will slowly pass through the block. Jello, tar, and even glass are also rigid fluids. (Medieval cathedral windows have, over the ages, become thin at the top and thick at the bottom because glass, being a fluid, flows down very slowly.) The flow speed or speed of an object passing through a rigid fluid must be less than the speed of vibration waves in the rigid fluid. The planets could pass freely through the æther because their speeds were much less than the speed of light. The planets could not move faster than light because, as the speed of light was approached, the æther would begin to act rigid rather than fluid. So the nineteenth-century æther theory anticipated what was to come.

Figure 1–18. Ice is a rigid fluid since the stone can slowly pass through it. Glass, jello, and tar are also rigid fluids.

The old æther idea also provided a picture of what electric charge and static electric fields were, and how they could produce electric force and waves. The æther even gave a vision of electrons and antielectrons (positrons), though they had not yet been discovered!

Suppose the æther in the vacuum of empty space is depicted as a gray void. Out of one area called *a* in the gray void you remove a bit of grayness, leaving a white area *b,* and you compress the removed gray bit into another gray area *c*. In this way you make *c* even grayer, or perhaps black. In effect, you have not created black or white, but have only separated black and white out of gray. So it is that + and − electric charges, called positrons and electrons, can be separated from the vacuum of space — in essence, dividing the void. It takes energy to do the division.

The energy is usually provided by a strong X ray (which is nothing but a very short light wave). The process, called pair creation, has been photographed many times.

The attraction between + and − charges was accounted for in the following way: If you take a piece of empty gray space and force the grayness to a different location, leaving white behind, you produce a strain or displacement in the space between the black and white areas. (Remember the æther was elastic.) That strain is the electric field, which was sometimes called a displacement field. If the plus and minus are not held apart, the strain will cause them to snap back together. (To this day some English physi-

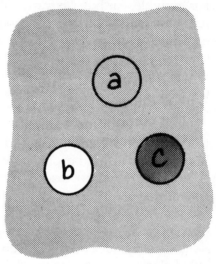

Figure 1–19.

cists call a strong electric field a high-tension field.)

If the plus and minus are allowed to snap together, you might expect that the strain would be relieved and everything would return to its prior state—a plain gray void. But when the strain in the earth is relieved by the snap of an earthquake fault, does everything go back quietly the way it was? Everything goes

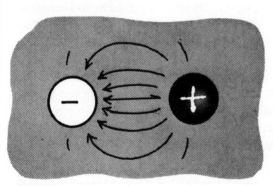

Figure 1–20.

back, but not quietly. The strain energy escapes, thereby making the surrounding earth shiver. So also the sudden relief of the strain between the plus and minus made the surrounding gray æther shiver. The shiver is the pulse of radiation always emitted by electron-positron annihilation. The radiation was visualized as the previously described shear wave running through the æther.

Incidentally, many of the old equations developed to describe the motion of waves in the æther are used today by earthquake geologists to describe the motion of earthquake waves in the earth's rigid crust. For example, earthquake waves, like light waves, can be polarized by reflection, and the æther equations describe the process. The æther equations actually fit earthquake waves better than æther waves because they describe compression as well as shear waves, and compression waves have been found to exist in the earth but not in the æther. (At one time, it was thought that X rays might be compression waves in the æther, but X rays can be polarized and compression waves can't be polarized.)

The old æther is bewitching and easy to fall in love with. But the hard historical fact is that, beyond the end of the nineteenth century, progress in physics was not made by those who fell in love with the æther. The turning point came with the theory and discovery of electrons. It had always been supposed that light moved slower in glass than in space because the æther was denser in the glass than in space. Then along came the idea that there were little electrically charged particles called electrons in the atoms. The light wave going through the glass would shake the electrons in the glass atoms. The shaking electrons would then send out new light waves. These new waves would interfere with the waves that entered the glass in such a way that the net effect of the whole business would be to reduce the wave speed inside the glass. So if there were electrons in the glass, the dense æther in the glass was no longer necessary. As time went on, one thing led to another. The electron idea explained all kinds of things that could never be explained before and eventually led to a whole new technology: electronics. The æther got in trouble with the Principle of Relativity and then with the Photon Theory. Gradually the æther faded into a haunting memory.

How the world is is so far
removed from how it ought to be
that anyone who proceeds to
reason according to how it ought
to be, rather than how it is, comes
inexorably to grief.

Chapter 2
The Central Problem

A Step Backward

It is commonly supposed that a discovery, particularly a purely scientific discovery, is always a step forward in understanding the world. But that is not so. The uncovering of a new phenomenon or the invention of an explanation for something previously unexplained evokes the joy of a step forward. But discovery sometimes shows a darker face. Sometimes a situation believed to be thoroughly explained, comprehended, and understood unravels, defying the supposed understanding. Joy is gone to grief and frustration. This chapter is a tale of the dark side of discovery.

The Problem

Light moves exceedingly fast. So fast that Galileo, after trying to time it, thought it must be infinitely fast.

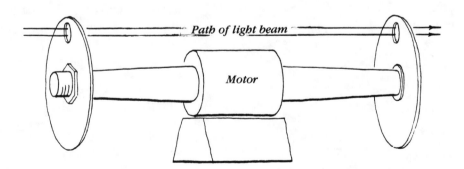

Figure 2–1. The speed of light could be measured by this super-simple device. The speed of light is determined by adjusting the speed of the motor so that the disks make one turn in the time it takes the light to travel the distance between the two holes.

Not until the year gold was discovered in California was the speed of light measured in a laboratory in Europe; it was found to be 670 million miles per hour. (Radio waves and X rays also travel at the speed of light.) That speed is ten thousand times faster than the earth's orbital speed around the sun. It is one million times faster than the speed of sound, which Galileo succeeded in measuring before he attempted to measure the speed of light. And it is ten million times faster than highway traffic. The speed referred to here is that of light through empty space or almost empty space—like air. In water, glass, or plastic the speed of light is cut, sometimes by as much as one-half.

Now suppose there is some kind of apparatus that measures the speed of the light passing through it. It reads 670 million miles per hour. Further suppose that the apparatus is put on a truck moving toward the light at 1 mile per hour. What would you expect it to read? Answer: 670 million plus 1 mile per hour. And

what if the apparatus is on a truck moving away from the light at 1 mile per hour? Wouldn't you expect it to read 670 million minus 1 mile per hour? Yes, that is what you would expect it ought to be. But astonishingly, that is not how it is. In spite of the trucks' speed, the apparatus in both trucks would still read 670 million miles per hour.

Why can't the speed of light be increased or decreased a little by moving toward or away from the light's source? In the year 1900 that was the most important question in the world. The speed of light's immunity from the rules governing the speed of everything else was shocking, embarrassing, and to some philosophers unbelievable.

Needless to say, various versions of the two-truck experiment were repeated many times before the scientific world became grudgingly resigned to the fact that running toward or away from a light beam, no matter how fast the run, has no effect on the perceived speed of the passing light.

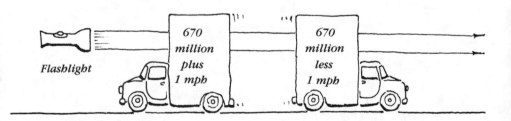

Figure 2–2. The two-truck experiment. Inside each truck is a device that measures the speed of light passing through the truck.

You can try supposing that light is composed of particles, or you can try waves. You can try supposing that Galileo's Dictum is valid or you can suppose it is in error. But try as you will, you still won't be able to make sense out of this situation. Why the speed of light is an absolute is an independent problem, indeed a paradox in and of itself.

Possibilities

This paradox could be attacked in two ways. One was to say that light is very weird stuff. In fact, it was unknown what it really was. (In fact, it is unknown what anything really is.) Some said light was waves, some said particles, and the Bible said it was just the absence of darkness.

Department of the Treasury
Internal Revenue
Service Center

FRESNO, CA 93888

29107   21

Date of This Notice
MAR. 19, 1979
Taxpayer Identifying Number

If you inquire about
your account, please
refer to these num-
bers or attach this
notice

WA   7910

578-52-0037   WX
Document Locator Number

LEWIS   EPSTEIN
% PHYSICS DEPT CITY COLLEGE
50 PHELAN ST
SAN FRANCISCO CA   94112

94254-457-55056-9
Form Number   Tax Period

1040   DEC. 31, 1976

1

STATEMENT OF ADJUSTMENT TO YOUR ACCOUNT

OVERPAYMENT ON ACCOUNT BEFORE ADJUSTMENT $268.00

ADJUSTMENT COMPUTATION

TAX- INCREASE   243.00
INTEREST CHARGED   26.33
NET ADJUSTMENT   CHARGE   269.33

BALANCE DUE $1.33

The numbers at the left identify the codes on the back
16,25   ◄———— of this notice that pro⸱⸱⸱ ⸱ ⸱her ⸱⸱⸱   ⸱⸱ ⸱tructions.

The IRS. It might be supposed that 670 million is such a large number that 670 million plus or minus 1 would still in effect be 670 million. But 670 million plus 1 is not 670 million. Practical proof: The United States government's Internal Revenue Service takes in much more than $670 million. Yet the IRS still went to the trouble of mailing a bill for an alleged underpayment of $1.33.

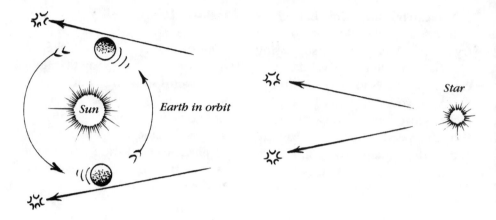

Figure 2–3. *The speed of starlight if the earth moves. The fastest object at our disposal is the planet on which we ride. The speed of starlight is measured as the earth moves toward and later away from a star. The measured speeds "should" differ by twice the earth's orbital velocity. But in fact they don't differ at all.*

Maybe something could be thought up that would always move past you in the same way, something you could not outrun. That is, something that remains fixed with respect to you, like a rainbow, which just steps right or left every time you step right or left.

Had I been a scientist in 1900, I would have tried to invent some weird thing that light could be that would explain why its speed always seemed to be the same. I would have reasoned, "If there is something strange about the speed of light, there must be something strange about light itself." But that was not the attack Einstein made.

A Wild Plan

Einstein said to himself, "I don't know what light is, and I don't care what it is. The problem is not with light; the problem is with speed." But what is speed? Miles per hour, feet per second, centimeters per centiday. Speed is a measure of space divided by a measure of time. So if the idea of speed is in trouble, it is because the underlying ideas of space and time need alteration.

I would like to warn would-be emulators of Einstein's approach that this kind of attack on a problem is in general foolish. If there is a problem with something, what is

$$Speed = space/time$$

usually wrong? The thing that can go wrong most easily. When the vacuum cleaner won't start, what is usually wrong? It is not plugged in. Only a flake would suspect that the coils in the electric motor have unraveled.

If a door in a house won't close, two things can be done. The door can be changed by planing or rehanging. Or the house can be changed by going down to the foundation with house jacks and jacking up the building until the door will close. Of course, if the one door is ever made to close by jacking, every other door and window in the house will jam. Jacking around with the foundation is usually a stupid approach.

Space and time are the foundations of physics. Space and time underlie every aspect of physics: mechanics, thermodynamics, electricity, and magnetism, as well as optics. But there was only a problem in optics, and only in one part of optics—light's speed. Suppose space and time could be jacked around to cure the speed-of-light problem. How could the new idea of space and time still square with all the other aspects of physics that previously worked quite well? Jacking around with space and time meant opening a Pandora's box of unforeseen consequences.

Even though the odds were powerfully against success, Einstein's attack carried the day. But it must always be borne in mind that the price of one successful attack of this type is ten thousand defeats. Just as the price of one new redwood is ten thousand seeds blowing in the wind. Perhaps that is why there are so many people on earth.

Review

Most speeds are relative. That means they depend jointly on the observer and the thing observed. For example, the relative speed of the wind can be increased or decreased by running against it or with it. But to everyone's amazement, light defies this expectation. Whether you run against it or with it, its speed remains 670 million miles per hour. The speed of light is not relative; it is absolute.

Einstein decided that the reason for light's unexpected behavior was some defect in the common concept of speed, hence in the concept of space and time.

Question

If a gun is fired forward from an onrushing locomotive, the bullet travels forward with the speed of the train added to the bullet's normal muzzle velocity. Is the speed of light that is fired forward from the headlight of an onrushing locomotive also increased by the locomotive's speed?

a The speed of light coming from

Figure 2–4.

the headlight is combined with the locomotive's speed and so is increased.

b The speed of light coming from the headlight of the onrushing locomotive is the same as the speed of light coming from a locomotive standing at a station.

Answer

The answer is **b**. Here's one way of reasoning it out: If you move toward (or away from) a source of light, your motion does not affect the speed with which the light passes you—recall the two-truck experiment (Figure 2–2). According to the Principle of Relativity, the outcome of any experiment must be the same if you move toward a thing or the thing moves toward you. Is this logic confirmed by actual experience?

As fast as locomotives move, they don't move fast enough to make a good test, but some stars do move fast enough. Perhaps the speed of light coming from a star moving toward the earth is faster than the speed of light from one moving away. Over one-third of all known stars have companion stars that orbit around them. Many of the companion stars alternately rush toward and away from the earth as they go around their orbits. Poincaré used these double stars to see if the speed of light depended on the speed of its source, and found the speed of light did not depend on the speed of the source. In essence, Poincaré's method of testing was very simple.

The orbiting companion of the illustrated double star is moving away from earth when it is at position *1* and toward the earth when it is at position *2*. If the speed of light

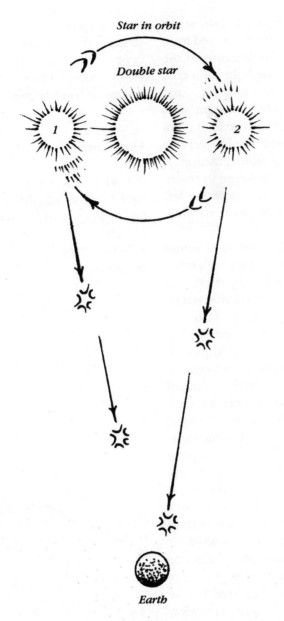

Star in orbit

Double star

Earth

Figure 2–5. *The speed of starlight if a star moves. Perhaps the speed of light coming from a star that's moving toward the earth is faster than the speed of light from one moving away. But tests show no difference. The orbiting companion of the illustrated double star alternately moves toward and away from the earth. If the speed of light depended on the speed of the star it came from, you might see the double star's companion arrive at position 2 before it left position 1. But in reality, you never do.*

depended on the speed of the star it came from, the light projected from position *2* would move toward the earth faster than the light projected from position *1*. In a long race, and the light coming from distant stars runs a very long race, the light coming from position *2* could overtake the light coming from position *1*. That means you would see the companion star arrive at *2* before it left *1*. But in reality you never see such things.

Incidentally, the speed of sound coming from the locomotive's whistle, like the speed of light coming from its headlight, is not affected by the locomotive's speed. (Do not confuse frequency with speed.) Yet, if you move toward a source of sound, the speed of sound relative to you is increased. Now you might ask, "How does sound get away with violating the Principle of Relativity? Must not all physics obey the same basic laws?" The answer is that in the sound story there is a third party that plays kingpin. It is the air. It could have been that there was something in space (like æther) that would do for light waves what air does for sound. But the world was not put together that way. Empty space contains nothing that can be used as a kingpin, that is, as a reference point to tell who is really moving.

And now for a final ironic twist. Even though the speed of the light coming from the onrushing headlight is not increased, the momentum and energy of the light it pro-jects forward is increased! How strange that the less familiar concepts of momentum and energy more closely obey intuition than the more familiar concept of speed.

Question

Can a light beam in empty space ever be observed moving obliquely (sideways), as illustrated in the first sketch?

a Sometimes.
b Never.

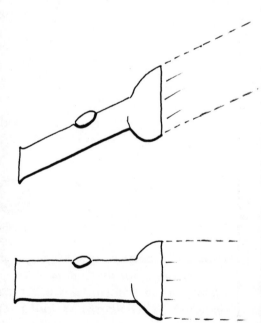

Figure 2–6.

Answer

The answer is **b**. This answer is based on the hypothesis that the light beam is a succession of waves. (Later on, I will show how the same answer can be arrived at on the hypothesis that the light beam is a succession of particles.) The waves move because each point on the wave spreads out into a little circular wave called a wavelet. All the little wavelets working together add up to remake the wave in a new position.

If the wave moves obliquely the circular wavelets must become elliptical. But they can't be elliptical. If the wave is going to spread with the same speed—the speed of light—in all directions, the wavelets must be circles and only circles.

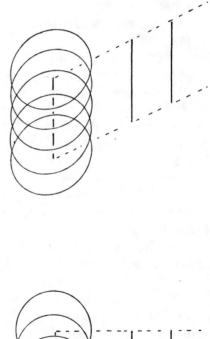

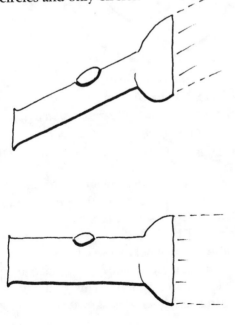

Figure 2-7.

Inside a physicist's head. The laboratory is the supreme court of physics, but the wheels of justice grind exceedingly slowly and theoretical physicists are impatient. Moreover, the two-truck experiment was almost impossible to do because of purely technical problems. Yet its result was widely anticipated theoretically. How was it foreseen? By understanding the rules of electricity and magnetism.

As confirmed by Hertz, light waves (and radio waves) are electromagnetic waves. They work like this: When a magnetic field increases or decreases in strength, an electric field is induced that curls around the changing magnetic field. This is exactly the idea behind an AC transformer. Conversely, when an electric field increases or decreases, a magnetic field is induced that curls around the changing electric field.

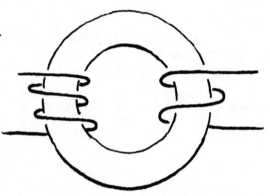

*Figure 2–8. Alternating current transformers work because the **changing** electric field in one wire coil makes a changing electric current in that coil, which makes a changing magnetic field in the iron ring, which makes a changing electric field curl around the iron ring, which makes a changing current flow through the other wire coil, which curls around the ring. But these fields can exist without wire coils and iron rings. (The metal helps; however, it is not essential.) In empty space a changing electric field directly makes a changing magnetic field, which directly makes another changing electric field.*

This duality is, in essence, the very heart of electromagnetic theory. It means that once introduced into the universe, an electric field or a magnetic field becomes immortal—it goes on some place forever. For if you discharge an electrically charged body and so try to kill the associated electric field, or if you destroy a magnet and so try to kill the associated magnetic field, the very death of either of the fields gives rise to a new field of the other kind: electric to magnetic or magnetic to electric. The dying field is a **changing field,** so it must give rise to a new field of the other kind.

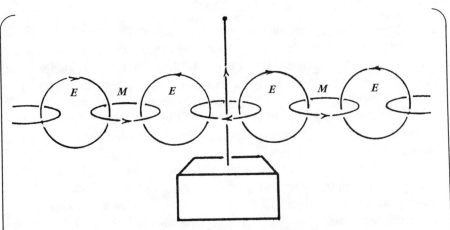

Figure 2–9. Changing fields induce other fields to curl around them. The alternating electric field in the transmitter antenna induces a magnetic field to curl around it. When, in turn, the magnetic field alternates, it induces an electric field to curl around it, and so on. The chain of curling fields makes the radio wave. Individual chain links do not change, they just move away from the antenna. As the links move, the field moves. Thus, the field at any fixed point in space continuously alternates as the links move by.

Note that there is no way to picture the chain links of an electromagnetic wave in less than three dimensions. A sound wave can be pictured in one or two dimensions. In fact, all the laws of mechanics can be illustrated on a two-dimensional pool table. But electromagnetic interactions are different. Unlike mechanics, the ideas of electromagnetism will not fit onto a two-dimensional space. Three-space dimensions are essential. Soon you will find ideas that will not even fit into three-dimensional space.

So on and on forever it goes. The electric field collapsing to make a magnetic field, the magnetic field collapsing to make an electric field—an electric field "reincarnated." This perpetual motion is the machinery that propagates radio waves, light waves, and even X-ray waves through space.

Travel and travel. Though the broadcasting station is long silent, the candle long out, the radiation lab long closed—the wave goes on, eternally faithful to its last command: **NEVER EVER STOP.**

Now suppose you are riding in a truck moving in the direction the wave is going. If you measure the wave as moving slower, then by riding faster you can measure the wave as moving even slower. By riding fast enough you can measure the wave as standing still.

If the wave is measured as stationary, the fields that make the wave are also measured as stationary or static. But you can't have static fields in empty space. To make static fields you need magnets and charged bodies. If you have magnets and charged bodies, space is hardly empty. And the static field is not independent of the magnet or charge that makes it. Destroy the magnet or discharge the bodies and the static field is destroyed.

Therefore, you can never perceive a light wave to be stationary in empty space. A light wave must move to exist because it is made of electric and magnetic fields, and these fields are self-sustaining only by the trade-off mechanism that depends on change. Seeing a static light wave would be like seeing an AC transformer operating on DC.

You might think your motion could allow you to perceive strange things, to perceive the laws of electricity and magnetism somehow changed. But remember that Galileo's Dictum and Poincaré's Principle of Relativity specifically forbid such changes, because such changes in the laws of electricity and magnetism, observed in a sealed laboratory, would be a tip-off that the sealed laboratory room was in motion.

To avoid any possibility of measuring a static light wave, you can't even be allowed to measure one that is slowed down (or speeded up), since repeated slow-downs would stop the wave. On the purely theoretical grounds just outlined, the result of the two-truck experiment was anticipated quite apart from and independent of any trial by experiment.

Intuition is science's most
powerful and yet most untrust-
worthy engine.

Chapter 3
Pandora's Box

Thought Experiments

"*Gedanken* experiment" is Einstein's German expression for a
thought experiment. It is an experiment you carry out in your head.
You can do it because you also have
in mind all the laws governing the
events you visualize. Einstein used
Gedanken experiments to figure
out what changes needed to be
made in his ideas of space and time
in order to make the speed of light
absolute rather than relative. He did
this by **reasoning backward.** In his
mind, he **supposed** that the speed
of light was absolute and he also
supposed Galileo's Dictum must always be true. Then he did *Gedanken*
experiments to see what consequences they would have on space
and time. Here is such a *Gedanken*
experiment; needless to say, its outcome will be surprising.

Synchronized Motion

Suppose a fleet of ships is spread
out in some formation. It is desired
that the whole fleet move without
changing its formation. To move
with an unchanged formation, all
the ships must move together. They
must all start simultaneously, all
slow down simultaneously, and all
stop simultaneously. If one moves a
little too soon or a little too late, the
distance between it and its neighbors will change and the formation
will be spoiled.

Let the fleet be three old NASA
spacecraft so far out in space that
you can forget gravity. A flag ship is
midway between the lead ship and
the rear ship. That way, orders transmitted from the flag ship will arrive
simultaneously at the lead and rear
ships. Orders are transmitted by
radio or signal flag; either way, the
message travels at the speed of light.

To begin with, the fleet is stationary—which, as Galileo would tell
you, doesn't mean it is standing still,
but only that it is moving at the
same speed you are; what that speed
is is quite unknown.

The fleet is going on maneuvers.
It is going to move in fixed formation. You are going to stay put and

watch. From the flag ship comes the order, "Burn for one minute on command...5, 4, 3, 2, 1, burn!" The command arrives at the lead and rear ships simultaneously, their engines fire simultaneously, and off goes the fleet. The distance between the lead and rear ships is unchanged. The flag ship adjusts its position as necessary to stay midway between the ships.

The whole fleet is now drifting slowly to the right. Yet, **from the crews' viewpoint they are all at rest** and you are drifting slowly to the left! Now here comes the first surprise.

same unchanging, absolute speed of light. The crews perceive you in motion and **themselves and their ships at rest, with the flag ship midway between the lead and rear ships.** So, as measured by the crews, the signal arrives at the lead and rear ships simultaneously. The crews perceive the whole fleet eating together. You perceive the rear ship's crew eating before the lead ship's crew.

It was this sort of *Gedanken* that forced Einstein to make the startling deduction that events happening at the same time for the fleet happen at different times for you, and vice

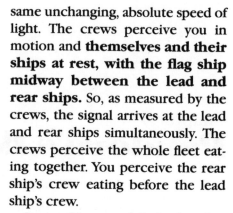

Figure 3–1.

Different Time

It is lunch time. From the flag ship comes the command, "Eat lunch." As measured by the crews, the order arrives simultaneously at the lead and rear ships. But as you measure it, the "eat lunch" signal must chase the lead ship, while it runs head on into the rear ship. You measure both the forward and rearward moving signals traveling relative to you at the same speed: the unchanging, absolute speed of light. So, as you measure it, the signal that has to chase the lead ship takes longer to arrive than the signal that runs head on into the rear ship. Einstein supposes the ships' crews also measure the signals traveling past them at the

versa. The old idea of time was that all parts of space arrive at the same time together, regardless of how anyone was or was not moving. But if the speed of light is to be absolute, then the idea of time must be changed, so that different parts of space do not arrive at the same time together! The price of making the speed of light absolute, rather than relative, is to make time relative, rather than absolute.

Relative time is the most essential new idea in the Special Theory of Relativity, and as you will soon appreciate, it opens a Pandora's box. Let's resume the thought experiments; you will see that the desynchronization of lunch time is not the most serious effect of relative time.

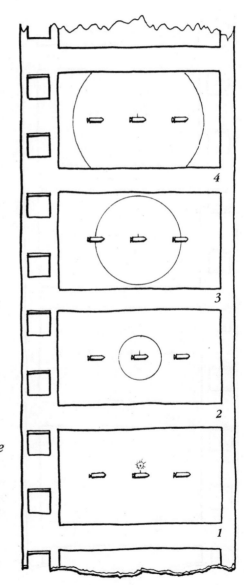

Figure 3-2. At rest. This film sequence shows the signal from the flag ship arriving at lead and rear ships simultaneously. When I use the expression "at rest," I mean at rest with respect to the observer. If you and the ships you observe are both flying through space together, that is, with the same velocity, then the ships are at rest as far as you are concerned.

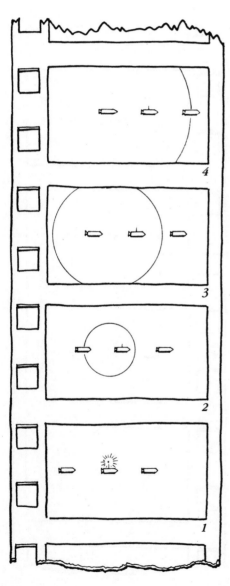

Figure 3–3. In motion you measure the signal arriving at the rear ship before the lead ship. If you do a Gedanken *experiment in which the flag ship simultaneously shoots a bullet toward the lead ship and toward the rear ship, and if you use the rules of the old physics in your* Gedanken *experiment, then the bullets will arrive simultaneously at the lead and rear ships as measured by you and the crews, even if the ships are in motion relative to you. The bullet that chased the lead ship had farther to fly than the one that ran head-on into the rear ship. But the speed of the forward bullet is the muzzle speed plus the flag ship speed. The speed of the rearward bullet is the muzzle speed minus the flag ship speed. However, the speed of light, unlike the speed of the bullets, is unaffected by the speed of the thing that shoots it.*

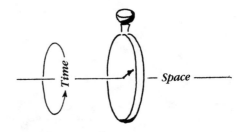

The thin clock reads time at one location in space.

Figure 3-4. A thick clock. The idea that all parts of space need not arrive at the same together is well visualized by a spinning cylinder. A stripe is painted along the side of the cylinder. If the cylinder is not sliding to the right or left, all parts of the stripe go over the top together. If the cylinder begins to slide, the stripe is perceived to wrap around the cylinder! So all parts can't go over the top together.

With a little imagination, you can, if you wish, think of the cylinder as a thick clock that extends through space. The stripe is the clock's hand. The wrapping gives the perception that different parts of the clock read different times.

The wrapping of the cylinder should suggest why the device used to measure the speed of light in the moving trucks gave an unexpected result. The twist in the shaft mis-aligned the timing holes.

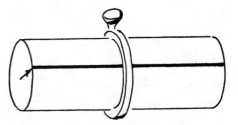

The thick clock reads time in different parts of space. If the clock is not moving, it reads the same time all over.

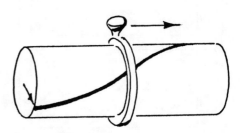

In motion the thick clock does not read the same time at all places.

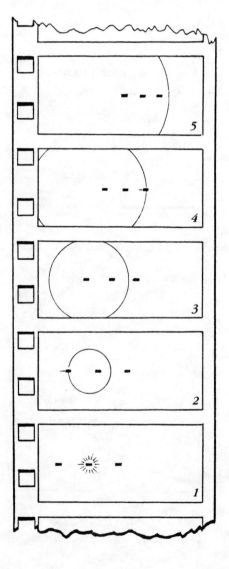

Figure 3–5. Separation between spacecraft shrinks as you measure it, as illustrated in this filmstrip, because the acceleration signal got to the rear ship before the lead ship.

Desynchronized Motion

Suppose the order for another engine burn comes from the flag ship. This will further increase the speed of the fleet relative to you. But as you measure it, which ship accelerates first? As you measure it, which ship receives its orders to accelerate first? The rear ship. As measured by the fleet's crew, the rear and lead ships receive the order simultaneously. As measured by you, the rear ship receives the order first and accelerates first. Thus, as you measure it, there is a time during which the rear ship is moving faster than the lead ship. During this time the space between the ships must decrease. When the lead ship finally receives its order, it too will accelerate. Then both ships will have the same speed and the space between the ships will stop shrinking.

So an immediate consequence of desynchronization is that you measure the space between the ships decreasing as they accelerate. The ships' crews measure their ships accelerating simultaneously and hence the space between the ships remains unchanged.

When the flag ship orders yet another acceleration, the space you measure between the ships will shrink again. And again, the shrink will not exist for the ships' crews.

When the flag ship orders deceleration, that order too, as you measure it, reaches the rear ship first. So the rear ship decelerates first. During the time interval between the

deceleration of the rear ship and the later deceleration of the lead ship, the space between the ships increases. When the fleet is completely decelerated, the space between the ships is back to what it was before the maneuvers began.

Now look away from this book for a moment and ask yourself how the space between the ships could decrease without the crews' knowing it. If the ships got close enough to each other, the nose cones would cram into the engine bells, so if the fools didn't see what was going on, they would certainly feel it!

Figure 3-6.

Figure 3-7.

No sideways contraction. It has been established that an object moving at high speed is perceived
to be contracted in its direction of motion. So you
should question if the object might also contract, or
even expand, in a direction perpendicular to its direction of motion. The answer is that the object's perpendicular dimensions will not be affected. Here is a logical argument
showing why.

Stationary

Moving

Suppose the moving object did contract in a direction perpendicular
to its direction of motion. And suppose the object was a bullet moving
down the barrel of a gun. Due to contraction the bullet would no
longer fit snugly in the barrel; the contraction would allow the expanding gas behind the bullet to blow by. Now reexamine this situation
as perceived by the bullet. The bullet measures its own "proper"*
diameter and measures the diameter of the gun barrel as contracted.
Thus, the bullet would jam and the gun would explode.

Now it might be that very high-speed bullets will allow blow-by, or it
might be that very high-speed bullets will jam, but it is logically impossible that both will happen. This argument can also be used against the
assumption that the moving object expands in a direction perpendicular to its direction of motion.

You might try to construct a logical argument, such as I have employed above, to show that an object cannot contract in its direction of
motion. Why don't you try? If you do, you will find you are always de-
feated, because there is time desynchronization in the direction of
motion. An object can be measured contracted in, **and only in,** its
direction of motion.

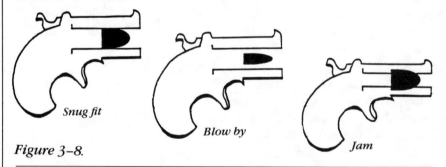

Snug fit

Blow by

Jam

Figure 3–8.

---

*"Proper" means the measure of a thing as perceived by an agent not in motion
relative to the thing being measured. If you ride on a ship, you measure its proper
length. If you measure the ship's length as it flies past you, you don't measure its proper
length. The idea of "proper" also extends to measures of time and mass.

Space Shrinks

The answer to this objection is that the ships are composed of many little atoms, each separated from its neighboring atoms by a certain space. The atoms are themselves little space ships. Just as you measure the space between the ships shrinking, so also you measure the space between the atoms shrinking. That means that the ships shrink in the same proportion as the space between them shrinks. Space itself shrinks, but it shrinks **only** in one direction, the direction of motion.

The effect is exactly **as if** you had a painting of the fleet and turned it sideways, so that everything in the painting shrank. The painting did not actually shrink. Nor did the crews on the ships in the painting (evoking artistic license) measure a change in their world (the world of canvas and paint). But if you perceive the painting obliquely, the crews in the painting must also perceive you obliquely. They would perceive you to be shrunk.

The painting was rotated in space to make the oblique view, so the painted fleet points in a different direction after rotation. The actual space fleet does not change its direction of motion through space, so it could not really be rotated in space. Yet it seems to be rotated and in fact is rotated, but not in space. I will return to this line of thought in Chapter 5.

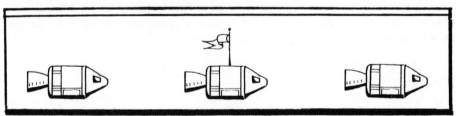

Figure 3-9.

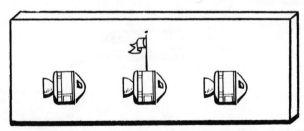

Figure 3-10. Is the shrinking of space an illusion? An illusion is usually something that can be seen by eye or camera but can't be touched or felt in any way, like a mirage. All things that see or feel the length of a thing at rest measure its proper length. All things that see or feel the length of a thing in motion measure its length shrunk. Would you call this an illusion or not? The word choice is up to you.

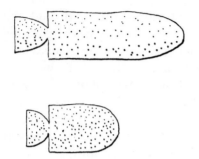

Figure 3–11. The ship shrinks in its direction of motion as you measure it, and you also measure its density as increased. Why? Because the atoms are crowded into a smaller volume. Additionally, the mass of each atom also increases, as will be explained in Chapter 7.

Epitome

To sum up and illustrate the new ideas developed in this chapter, consider how you, standing by a runway, would measure a spacecraft making a very high-speed landing.

If the spaceship pilot lowers both landing skis simultaneously, you measure the rear ski lowered first. If you measure both skis hitting the ground simultaneously, the pilot measures the nose ski hitting first. If you measure the ship landing in a horizontal attitude, the pilot measures it landing in a nose-down attitude—that is the only way he can measure the nose ski hitting first. All this is a consequence of time desynchronization. You measure the ship's length to be shrunk. The pilot measures the landing field's length to be shrunk. Everyone measures his own proper length. This is also an indirect consequence of time desynchronization.

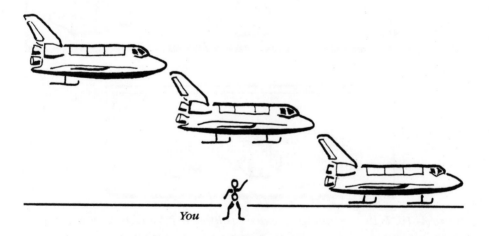

You

Figure 3–12. You measure the rear ski lowered first for the same reason you measure the rear ship's crew eating lunch first. For the touch-down situation it is just vice versa—you measure two events as simultaneous, so they can't be simultaneous for the moving crew.

Question

You measure a distant star directly overhead. Suppose the star is pulsing, and the light pulses from the star fall flat onto the runway. If you go on board the ship and drive it along the runway as a rocket sled, you will measure the light pulse as hitting the runway at

a the tail end of the ship before the nose end.

b the nose end before the tail end.

c both ends simultaneously. that is, falling flat.

Figure 3-13.

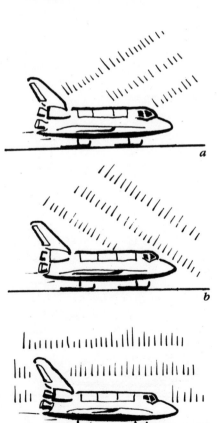

a

b

c

Figure 3-14

Answer

The answer is **b.** As measured by anyone standing on the runway, the pulse will hit the runway at the nose and tail ends simultaneously. As measured by anyone inside the moving ship, the pulse will hit the runway at the nose end first. So the pulse must be tilted. This is just like the spacecraft touchdown situation described in the epitome. Because the pulse is tilted, it must seem to those on the ship that the star is no longer overhead, but is slightly forward. Strange? Recall your experience with rain falling straight down.

If you get on your bicycle and ride, is it still coming straight down?

This is a very real effect. As a spacecraft changes its speed or direction of motion, passengers on the ship see all the stars change their positions in space.

When the ship is moving very rapidly, all the starlight seems to be coming from the forward direction, and the sky to the rear of the ship may become devoid of stars. When your bike is moving very rapidly, all the rain seems to be coming from the forward direction, and your rear side stays dry.

In particular, as spaceship earth circles the sun, all the stars appear to move in tiny circles. How long does it take the stars to go around their tiny circles? One year. It takes careful telescopic observation to see this effect, called the "aberration of starlight," but it was detected in 1726, well before Einstein's grandfather was born.

This aberration effect has nothing to do with parallax. The "parallax effect," first detected in 1836, depends on the distance of a star. The aberration effect is the same for all stars.

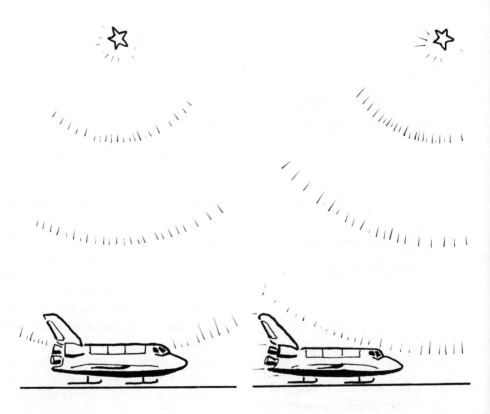

Figure 3–15.

Figure 3–16.

Figure 3–17. *Floating at rest in space, the stars seem to spread all around the sky. But as seen from a speeding spacecraft, those same stars are not spread all around. They are concentrated in the part of the sky toward which the observer's craft is moving. The concentration superficially resembles a giant globular cluster. By looking at the sky, can the observer know he is in motion through space? No. The observer only knows he is in motion* **relative** *to the stars. Note: The stars are presumed to be very remote from the spacecraft, so the business described here has nothing to do with stars streaming by the craft or other ordinary three-dimensional perspective effects.*

Question

You are inside a spacecraft that is a mile long. An alien spacecraft flies past you, or perhaps you fly past the alien (there is no way to distinguish between these possibilities). When the passing alien is alongside you, you note that the alien craft is exactly as long as your own craft. That is, the alien nose is abreast of your tail when the alien tail is abreast of your nose—according to your time. Are the two craft identical in length?

a Yes, they are identical in length.

b No, they are not identical in length.

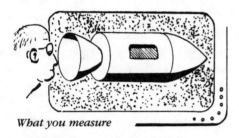

What you measure

Figure 3–18.

Answer

The answer is **b**. The craft are **not** identical in length even though you measure them to be equally long, because you measure your craft at rest (since you are riding inside it), but you measure the alien in motion and a thing in motion is measured as being foreshortened. Your craft is properly a mile long, and the alien craft is properly more than a mile long. Proper length is the length of a craft as measured by its occupants.

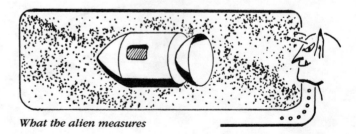

Figure 3–19.   *What the alien measures*

Question

You measured the passing alien craft to be exactly as long as your craft. Let's consider what the alien measures:
a The alien measures its craft to be as long as yours.
b The alien measures its craft to be longer than yours.
c The alien measures its craft to be shorter than yours.

Answer

The answer is **b**. If you measure the moving alien craft to be a mile long, then its proper length is more than a mile and the alien riding inside the alien craft measures its proper length. The alien observes your craft in motion. If your craft is properly a mile long, and the alien measures its length in motion, the alien measures it to be less than a mile long.

But if you measure the alien nose, which is abreast of your tail, at the same time the alien tail is abreast of your nose, doesn't the alien measure the same coincidence of events? No. Why not? Because these events happened at the same time according to your time. Events that happen at the same time for you don't happen at the same time for the alien. The desynchronization of time is the root cause of the foreshortening effect.

What if your craft and the alien craft were really identical, that is, if each were properly a mile long? Then when you passed each other, each of you would measure the passssing craft to be less than a mile long.

Question

Again return to the situation where you measure the passing alien craft to be exactly as long as your craft. You, riding in the nose of your craft, time how long it takes you to fly past the alien craft; and the alien pilot, riding in the nose of its craft, times how long it takes it to fly past your craft.
a You take as much time as the alien.
b The alien takes more time.
c You take more time.

Answer

The answer is **c**. The speed of the alien relative to you is equal and opposite to your speed relative to the alien. But the proper length of the alien craft is greater than your craft's proper length. Passing time is proportional to proper length. Remember, you measure the alien to be a mile long. The alien measures you to be less than a mile long, so of course it takes you more time to pass the alien than for the alien to pass you.

At the beginning of the next chapter this kind of situation will be reconsidered in detail, with one essential difference: The two passing craft will be properly identical, that is, each will be a mile long as measured by its occupants.

Question

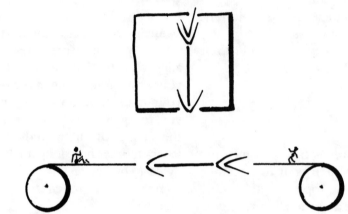

Figure 3–20.

A square box is falling on a conveyor belt, as illustrated. How will the box be measured by aliens who are riding the belt?

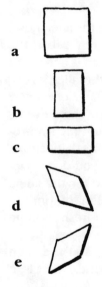

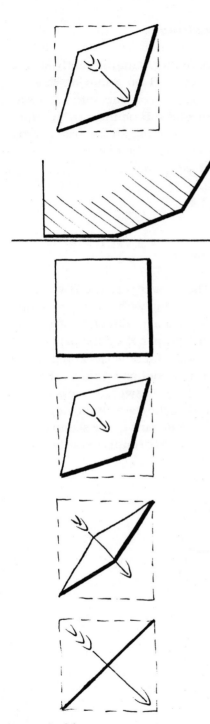

Figure 3-21.

Answer

The answer is **e**. The belt is moving left, so the aliens see the box moving down and to the right. To make life simple, assume its speed to the right is the same as its speed downward. It is measured to move at a 45-degree angle, as illustrated. The box is foreshortened in, and only in, its direction of motion, as illustrated. Note that the aliens must therefore measure the left corner of the box as hitting the belt before the right corner. You, who are not riding on the belt, measure both corners as hitting simultaneously. Foreshortening always requires desynchronization, and vice versa.

Also note that as the bottom of the box hits the belt, the aliens will measure a "kink deformation" in it. The part of the box to the right of the kink will still be falling; the part to the left will not. You will measure no kink and all parts of the bottom will stop at once. Though the box may be rigid, it is not perceived as rigid by the aliens.

As the measured speed of the box increases, the box is measured as more flattened in its direction of motion. In the limiting case, as the speed approaches the speed of light, all lines in the box become perpendicular to the box's direction of motion. Not only light waves, but anything that has line structure in it and moves at the speed of light, must have that structure perpendicular to the direction of motion.

Question

A rubber stamp instantly stamps the letter **H** on a moving sheet of paper. So you measure an exact copy of the **H** on the moving paper. An alien riding on the paper measures the **H** to be like

a **H** as you perceive it.
b Ⱨ.
c **H**.

Answer

The answer is **c**. The **H** you measure moving with the paper is foreshortened in its direction of motion. So the proper **H** on the paper must be wider. But how can a standard **H** on the rubber pad print an **H** wider than itself? Answer: As measured by the alien, the stamp is moving to the left and all parts of the stamp do not hit the paper simultaneously.

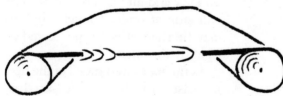

Figure 3–22.

Playing with time offsets. Pretend that you could make clocks in different locations read different times, for example, make one clock run one hour behind the other. And pretend that you could fake out the people who use those clocks so they would suppose the clocks were not set incorrectly (offset). You could deceive and cheat those people in many ways. Pause for a moment and think about how the cheating would work.

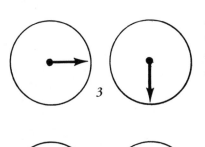

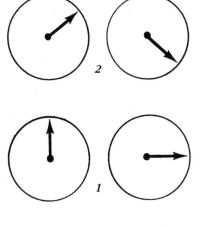

Figure 3–23. These clocks are offset or desynchronized. There is a time difference between them—three hours—but that difference does not increase or decrease with time.

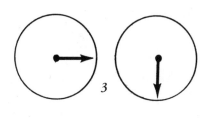

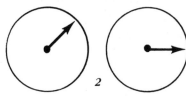

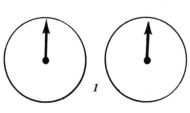

Figure 3–24. These clocks are running at different rates or speeds. To begin with there may be no time difference between them, but gradually one runs ahead of the other and the time difference between them continuously increases with time.

Question

To make a runner's time seem better than it is, the finish clock should be set to run
a slightly behind the starting clock.
b in synchronization with the starting clock.
c slightly ahead of the starting clock.

Answer

The answer is **a**. The **a** clock is set back so that it shows less time spent in running. Other ways to cheat would be to slow down the clocks or reduce the running distance.

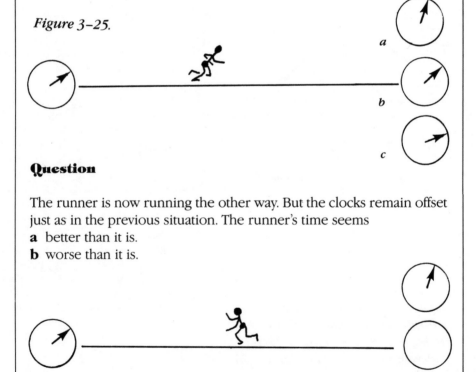

Figure 3–25.

Question

The runner is now running the other way. But the clocks remain offset just as in the previous situation. The runner's time seems
a better than it is.
b worse than it is.

Figure 3–26.

Answer

The answer is **b**. Since the runner is going the other way, the starting clock becomes the finish clock. The finish clock is now set ahead of the starting clock so it shows extra time. Desynchronizing the clocks helps when running one way and hurts when running the other. Tampering with the speed of the clocks or the running distance would have the same effect in both directions.

The measured speed of things can be easily altered by offsetting time. Here is an important example. You measure a light flash taking a certain time to fly the length of a meter stick. If the stick is mounted on an alien planet that is moving past you in the same direction the flash is going, the flash takes more time to fly from one end of the stick to the other. If the alien planet is moving against the flash, the flash requires less time to fly the length of the stick.

Could the aliens' clock be set incorrectly to fool them into measuring both flights as taking the same amount of time? How? The trick is to mis-set the alien clock that times the start of the meter flight relative to the alien clock that times the end of the flight. **But remember, the aliens are faked out.** The aliens suppose their clocks at the ends of the stick are synchronized, that is, not offset. You measure the alien clock at the trailing end of the stick as reading ahead of the one at the leading end. How much ahead depends on the length of the stick and the speed of the alien planet. The aliens are faked out; regardless of how long the flight takes as you measure it, for the aliens, using their alien offset timing, it always takes the same amount of time.

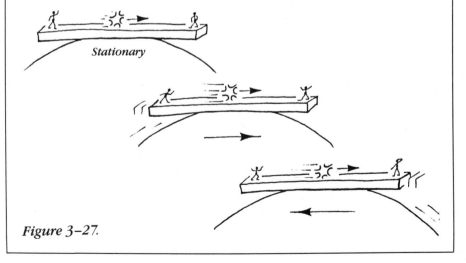

Stationary

Figure 3–27.

Seeing or measuring? I often use the words "you measure" or "the alien measures." Why not use the words "you see" or "the aliens see"? Because what you see is wrong.

For example, you may see two stars very near each other in the sky, yet they may not be at all nearby in space. One may be far behind the other, and so the apparently brighter star may actually be fainter.

Suppose both stars explode tonight. You will not see them explode tonight. It will be years before the flashes reach earth. And when they arrive they will not arrive together. The flash from the nearer star will arrive first. Though the stars explode at the same time, you don't see them explode together. Is this the celebrated time desynchronization of relativity? Certainly not. This kind of desynchronization was thoroughly understood three centuries before Einstein was born.

Signal transmission delay time allows you to see, hear, and do all kinds of things incorrectly. Because of transmission delay time, the battle of New Orleans was fought after the War of 1812 was over!

Suppose a pair of stars (or wheels) is moving side by side through space. You see the more distant star (or wheel) further in the past, and so it appears to lag behind the nearer star. Because of this effect, a box (or asteroid) moving rapidly through space appears to be turned.

If a pulsating star is moving away from you, the signal transmission delay time increases between each pulse. Each pulse is delayed more than the one before it. So the time between pulses appears to be longer than it properly is. If the star moves toward you, the effect is reversed. But don't confuse this with Einstein's slow time; his slow time does not depend on signal delay time or on the direction of motion.

Signal transmission delay time

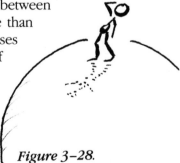

Figure 3-28.

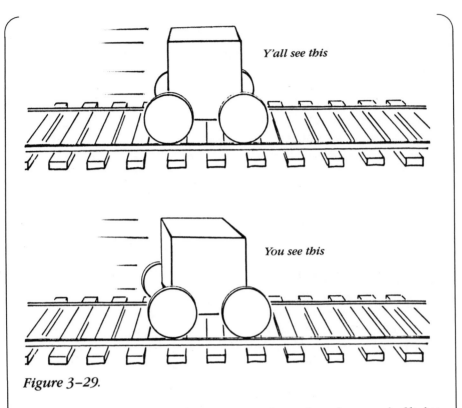

Figure 3–29.

can even cause you to **see** things moving faster than the speed of light! Suppose a gun, one light year away from you, is fired directly toward you. It will take one year for the first light from the oncoming bullet or gun flash to hit you. If the bullet moves at three-quarters the speed of light, it will hit you four-thirds of a year (a year and four months) after it is fired. The last light from the bullet will hit you just before the bullet hits you. As you see it, the time between first and last light from the bullet is four months. During those four months you will **see** the bullet travel one light year. So the bullet **appears** to you to be moving three times faster than light.

Everyone knows about and allows for all these signal transmission delay time effects. There is nothing new here. There are two ways to dispose of them: (1) Subtract the delay time from the apparent time, or, easier yet, (2) get so close to the happening that you can forget about the delay. If you are close enough to the lightning, the thunder is not delayed.

Sometimes a happening involves two places, like a starting line and a finishing line. How can you be very close to both? You can't. So you must have a team of assistants called local observers. Everyone watches what he is assigned to watch up close, thereby eliminating delay time. After the happening, all assistants mail in their notes and you work out exactly what happened. You make your measurements from the collection of notes.

The words "you see" should be replaced by the words "your local observers see" or "y'all see." It is required that, while y'all might be spread out in space, y'all must travel through space together—y'all must hold your formation. Were that not required, y'all could never agree on the speed of passing objects.

I use the word "measure" because it suggests the possession of more information than just your lone vision of distant events. I use the words "you measure" as a synonym for "y'all see."

Figure 3–30. The Doppler effect. Think of a bug sending out little waves as it walks on the surface of water. The waves are circles. The largest circle is the wave that has had the most time to expand. The center of the large circle is where the bug was when it started that wave. Since then the bug has moved. Can you tell which way it is moving? If the bug became a star and the ripples became light waves, the picture would hardly change.

Chapter 4
Measuring the Consequences

Two Kinds of Speed

Once a change in the way time works is introduced, many conventional, commonsense expectations must be altered. Indeed, Pandora's box is now open, and your ideas about how the world works will be shaken.

Suppose that two identical spacecraft fly past each other and you want to find the speed with which they are passing each other—their relative speed. There are two ways to do it. Way A: Riding in the nose of your ship, see how long it takes you to pass the other ship, that is, to go

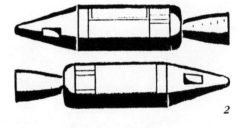

2

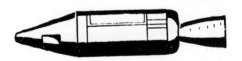

1

Figure 4–1. According to the old conventional ideas of space and time, you would expect the nose of your ship to reach the tail of the alien at the same moment the nose of the alien ship reached your tail, regardless of how things were timed and regardless of who did the measuring— provided only that the two ships had equal proper length. Note: The time sequence in this illustration runs from bottom to top, like the films.

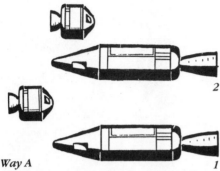

Way A   1

Figure 4-2. Because you measure the length of the alien ship as shrunk, your nose takes less time to pass it than the old conventional ideas of space and time would lead you to expect. What about the alien nose? How much time does the alien pilot riding in the nose of his ship and using his clock take to pass your ship? This question seems to bring common sense into conflict with the principle of relativity.

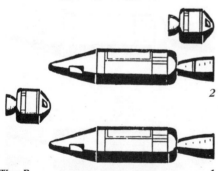

Way B   1

Figure 4-3. If you and your tail agent measure how much time the alien nose requires to pass your ship, that measurement will not be affected by the reduced length of the passing alien ship. Does the alien pilot using his clock measure the same amount of time spent to get his nose past your ship?

from the nose to the tail of the passing ship. Way B: See how long it takes the nose of the passing ship to go from the nose to the tail of your ship. Note that this method requires you to have a local agent in both nose and tail of your ship. The two ways should each lead to the same result; but they don't. They don't if you have jacked around with your idea of space and time.

The time measured by Way A is shorter than the time measured by Way B. How come? Because the passing ship, as you measure it, has shrunk. So as your passing speed goes up, you pass the ship more quickly, for two reasons. First, you go faster, and second, the thing you go by is shorter.

The time it takes the nose of the passing ship to go from nose to tail of your ship is also shorter, but only because the nose goes by faster. The length of the passing ship does not matter in Way B.

Now comes the pinch. An alien riding in the nose of the passing ship decides to see how long it will take him to pass your ship. What should that time be? The same time you found it took the nose of your ship to pass the length of his ship? Or the time you found it took the nose of his ship to pass the length of your ship?

This is not an easy question. The key to the answer is that the alien must measure your ship exactly as you measure his ship. No one has a special view; everyone has the same view. That is the essence of the Principle of Relativity. So the answer must be A, not B.

But that means that the alien finds it takes him less time to pass from the nose to the tail of your ship than you and your agent find it takes him to pass that same distance. How can that be? It can only be by accepting another basic change in the way time works.

It must be that you measure the clock carried by the alien to be ticking slow. The moving clock must be measured as slowed by exactly the same factor that the moving spacecraft is measured as shrunk.

This means that if you zoom around a track and carry a stopwatch with you, your time will be better than if you just leave the watch at the finish line, because your speed makes the clock go slow! By carrying the watch you get what is called **proper speed.** If the watch stays at home, you get what is called **coordinate speed.**

A locomotive engineer timing the passage of mileposts with his pocket watch measures the train's proper speed. Speed-of-light measurements always give the light's coordinate speed, because the timing device cannot ride with the light. After reading the next chapter, ask yourself what the proper speed of light would be if it could be measured.

The cause of time desynchronization and space shrinking was directly visualized. I will now show that it is even easier to visualize why moving clocks are perceived to run slow.

Figure 4–4. The locomotive crew determines the proper speed of the train by counting the mileposts passed per minute of time as measured by the engineer's pocket watch. If the station master's watch were used to time the run over a known length of track, the train's coordinate speed would be determined.

Figure 4-5. Stationary light clock.

Slow Time

To see the direct cause, you must look into the mechanism of a clock. At the heart of every clock is something that goes through some repeated motion: a pendulum, a balance wheel, a vibrating quartz crystal, or a vibrating tuning fork. Now in this story the bad boy, the troublemaker, is light. You want to keep your eye on him. So let light be the moving thing in the clock.

Let the clock be an empty tube with a mirror at each end. A flash of light is trapped inside and it bounces back and forth forever. Suppose each bounce takes 1 second. The tube would have to be 186,000 miles long and the mirrors 100 percent reflecting. So the light clock is hardly a practical timepiece, but this is a *Gedanken* experiment, and for *Gedanken* experiments equipment is engineered for conceptual economy.

How will a light clock be perceived by aliens riding by it in their ship? The aliens will measure the light clock as drifting backward by their ship. As it drifts by, the trapped flash goes up and down following a diagonal path.

But the distance between its bounces, measured along the diagonal, is longer than the distance between bounces if the flash simply goes straight up and down. When the flash goes straight up and down, it takes 1 second to bounce, because the clock was made to work that way. Since the aliens measure the

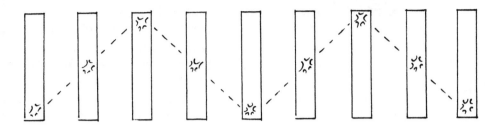

flash as taking a longer path between bounces, AND SINCE THE SPEED OF LIGHT CANNOT INCREASE, the aliens must measure more than 1 second between bounces. The aliens measure your light clock as running slow. They measure more than 1 second between your seconds! The aliens measure your time to be running slow.

Now hold on. Just because some imaginary light clock is measured as slowing down, how can it be concluded that time itself is perceived to slow down? It can be said that time slows down if every clock slows down. But there are all kinds of clocks. And indeed all slow, and by exactly the same amount as the light clock slows. For example, take your heart. You sit on the light clock. Suppose each time the flash bounces your heart beats. The light clock might even be your heart's pacemaker. The aliens must also see one beat for each bounce. So the aliens must measure your heart as slowing exactly as much as the light clock slows. You measure your heart, your light clock, and indeed all your clocks as running at their proper speed, regardless of what the aliens measure.

Figure 4–6. Moving light clock. The vibrating thing in ANY clock must be measured as vibrating slower as the clock moves faster, and the vibrating thing inside the clock must stop vibrating when the clock moves at the speed of light. Because if the clock were moving at the speed of light and the thing inside were still vibrating, then the vibrating thing's total speed through space would have to be faster than light.

Does the photon know that the light clock is moving?

The photon's motion in the light clock resembles the motion of a gymnast bouncing on a trampoline mounted on a traveling truck—except that the photon's speed is abruptly reversed at the top mirror, while the gymnast's speed is gradually reversed by gravity. However, and this is the key point, if the truck travels smoothly, without changing speed or accelerating, that is: no starting, no stopping, no jerking, no turning, then the gymnast is completely unaffected by the truck's travel.

Although you measure the gymnast as traveling sideways, as well as bouncing up and down, the gymnast and truck driver measure only the up and down part. So, as you measure it, the gymnast is moving a bit faster than the truck driver measures it. As far as the gymnast is concerned, the truck might as well be stationary. Moreover, a trampoline standing on bedrock, measured from the sun, is traveling, due to the earth's journey through space. Yet even though the earth travels through space faster than the gymnast can bounce, the gymnast is never concerned by or even aware of the ground's motion because it has no effect on the acrobatics. So also the photon is unaware of the light clock's motion.

In essence, this is the old Theory of Relativity—Galileo's Dictum about the relativity of space: Anyone, no matter how he travels, just as long as he does not accelerate, perceives himself fixed in space. You will come to appreciate that the new Theory of Relativity in essence is this old theory of the relativity of space coupled to the relativity of time.

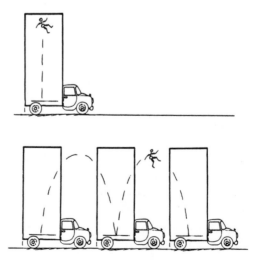

Figure 4-7.

The reason the aliens measure your time as running slow is because they perceive your light clock to be drifting through space. But you likewise perceive the aliens as drifting through space. So everything that holds for your clocks also holds for the aliens' clocks. You measure the aliens' time as running slow—exactly as slow as they perceive your time.

This is quite strange indeed. First of all, it is strange that time should be measured to go slow. But, accepting that, it would seem as if you should perceive my clock to be running fast (not slow) if I perceive yours to be running slow. Just as I become a giant if you become a midget. But it does not always work that way. Suppose I walk away from you into the distance. As I see you, you become a midget. But as you see me, I do not become a giant; I too become a midget. Artists explain that this works by three-dimensional perspective. The time effect might (as you will see in Chapter 5) be explained by making time a dimension and going to four-dimensional perspective.

Cosmic Speedometer

The light clock reveals that you must measure time for a moving thing as being slowed down, and it also enables you to visualize exactly how much it is slowed.

Suppose a photon in the stationary light clock can pass from P to A in 1 second. Now, rather than you drifting past the light clock, suppose (which must amount to the same thing) that the light clock travels past you. The clock travels from P to Q in 1 second. It might be expected that the photon would still go from bottom to top, that is, from P to D, in 1 second. But the distance $\overline{PD}$ is longer than the distance $\overline{PA}$.

There is no way the photon, traveling at the speed of light, can be perceived to make it from P to D in 1 second. So where is the photon measured as being in 1 second? The answer can be logically deduced in this way: First, the photon must be somewhere in the clock, that is, somewhere between D and Q.* Second, the photon always travels at the speed of light, so after 1 second its distance from P must correspond to the distance $\overline{PA}$. Draw a circle with the radius $\overline{PA}$ centered on P. After 1 second the photon must be somewhere on that circle. The point at which that circle intersects the line $\overline{DQ}$ satisfies both the first and second conditions. The photon must be measured at that point, called B, 1 second after leaving P.

If the clock traveled from P to C in 1 second, it would be traveling at the speed of light. The photon would have to use all its speed just to keep up with the clock. There

---

*When the clock moves, why doesn't the photon just hit the side of the light clock tube? Because the photon does not know the clock is moving.

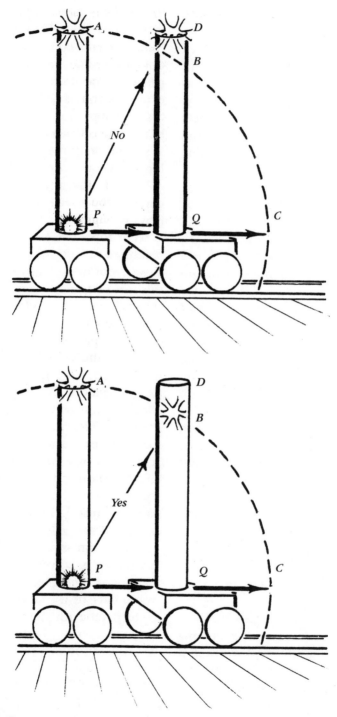

Figure 4–8.

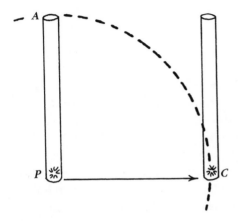

Figure 4–9. Light clock traveling at the speed of light.

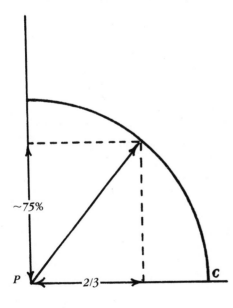

Figure 4–10.

would be no speed at all left for upward movement. The photon would just go from *P* to *C*, so after 1 second had elapsed the photon would still be at the bottom of the clock. The clock, as you perceive it, would not be running.

Now *Q* is midway between *P* and *C*, so the clock that travels from *P* to *Q* in 1 second is traveling at half the speed of light. By measuring the sketch, you will find that the photon at *B* is roughly five-sixths of the way to the top of the clock. So, even though 1 second has passed for you, only ⅚ of a second has passed for the traveling clock. Eventually, the photon will get to the top of the traveling clock, and then 1 second will have passed for that clock. You measure the clock traveling at half the speed of light as ticking off approximately 50 seconds, while your stationary clock ticks off 60 seconds. This relativistic effect is not very pronounced, even at half the speed of light; small wonder it escaped notice until the twentieth century.

Any clock problem can now be solved. Draw a cosmic speedometer, which is simply a quarter circle. If the clock is traveling at, say, two-thirds the speed of light, mark off two-thirds of the bottom side and then tip the needle of the cosmic speedometer until it lies two-thirds of the way over, as illustrated. The tipped needle now reaches only approximately 75 percent of the way to the top of the circle. This means that a clock traveling at two-thirds

The æther and who is really moving. It is easy for you to see that it does not matter if you drift by a thing or the thing drifts by you. It was also easy for Galileo to see it—though not as easy as for you. But it was not easy for Einstein or people of his generation to see it. Why? Because his generation lived in the shadow of the æther.*

The æther was a mysterious rigid but fluid substance that was supposed to fill all empty space. Rigid so it could transmit transverse light waves, just as the earth transmits transverse earthquake waves. Fluid so the planets could pass through it.

It seemed to that generation of scientists that if light was a wave in the æther, and the light clock was moving through the æther, that would affect the way the wave moved relative to the clock. So it would make a difference if you moved or the clock moved.

Einstein threw all the æther ideas out of his head and went back to Galileo's Dictum. It was not easy. According to the dictum, everyone measures the light clock riding with them as acting properly and everyone measures clocks in motion relative to them either as not affected or as all affected in the same way. So there is no way to tell who is doing the moving.

---

*ÆTHER = $\alpha\iota\theta\eta\rho$, derived from $\alpha\iota\theta\epsilon\omega$, a Greek word that means something like "I burn" or "I am in eternal motion."

the speed of light is measured as running at 75 percent or three-fourths of its normal rate. But don't forget that anyone riding with the traveling clock perceives the clock as running at its normal rate and your clock as running at 75 percent of normal.

The cosmic speedometer shows how you measure a clock as ticking slower and slower as the clock travels faster and faster, until finally, when the clock travels at the speed of light, the clock stops ticking. What happens if the clock travels faster than the speed of light? It can't. Why can't it? Because it would outrun the flash. So the flash could not remain in the light clock tube. This would tell anyone sealed inside the clock that it must be moving. But if someone sealed inside can detect motion, Galileo's Dictum is violated, and that cannot be.

Question

How fast does a clock have to travel in order that its rate is measured as cut by 50 percent, that is, it runs at half speed?

Answer

Tip the cosmic speedometer needle until it is halfway down from the top. Then measure how far it lies over. You will find it lies approximately five-sixths of the way over. To cut the clock's measured speed

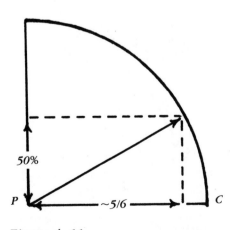

50%

P   ~5/6   C

Figure 4–11.

in half, the clock has to travel at five-sixths of the speed of light. You are now getting the feel of relativity.

Light Waves

Thus far, according to the oldest and simplest notion of light, I have pictured individual particles of light, which in recent times have come to be called photons. A flash of light must be a front composed of a swarm of these particles. The wave theory of light depicts these swarms as the waves. I would like to show you some things about these swarms.

To keep it simple, let there be one row of photons in the swarm or wave. As you see them, the photons are moving up, as illustrated in Figure 4–13, Sketch 1. But how will they look to someone moving to the left? Well, they will seem to be moving up and a little to the right. But how? Will they seem to be moving obliquely to the right, as shown in Sketch 2, or directly to the right, as shown in Sketch 3?

The answer must be 3, since light waves can't move obliquely. Why can't they move obliquely? Because all the photons started from the bulb when it flashed. Thus, they must all be on a circle centered on the position occupied by the bulb when it flashed. Sketch 3 satisfies this condition; Sketch 2 does not. In fact, each photon in Sketch 2 has a different speed, since each has traveled a different distance from the flash location, and that cannot be.

Figure 4–12. Just as a continuous line is thought of as a succession of points, so also can a continuous light be thought of as a succession of flashes.

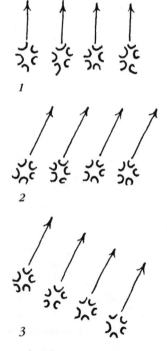

Figure 4–13.

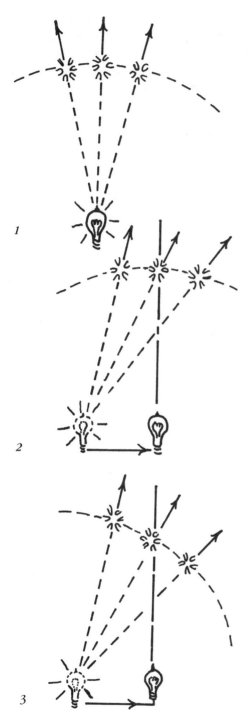

Figure 4–14.

What about the spaces between the photons? As the row of photons tilts will the spaces expand, contract, or remain constant? The spaces remain constant because things shrink in, and only in, their direction of motion. The distance between these photons is measured perpendicular to their direction of motion. The measured perpendicular separation between objects is always the proper separation between the objects.

Desynchronized Time

Now let's apply these ideas to the light clock. Up the light clock there goes not one photon, but a swarm of photons forming a wave. A person at rest with the clock, that is, riding with it, always measures the wave as going up the clock in the proper way, that is, the wave is perpendicular to the clock's axis. However, if the clock is perceived in motion, the wave must be tilted as illustrated. So one side of the wave reaches the top before the other! This is precisely the desynchronization of time already discussed. Desynchronization showing up in the small diameter of the light clock itself!

But that isn't all. If the wave in the stationary clock reaches from wall to wall, so must the wave in the moving clock also reach from wall to wall. But how can it reach if it is tilted? By stretching? No. Remember that the space between photons

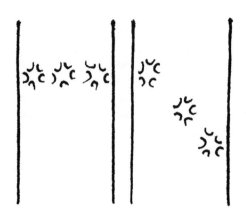

can't change. The only way the wave can reach is if the clock itself is measured as having shrunk in its direction of motion, and this shrinking has also already been described.

So now you have a super bonus. Careful drawing of the light clock will not only show by how much time is slowed at a given speed, but it will also show by how much time is desynchronized and by how much things are measured as having shrunk.

Figure 4–15. *This sketch shows the tilted row of photons stretched, but the row of photons can't stretch.*

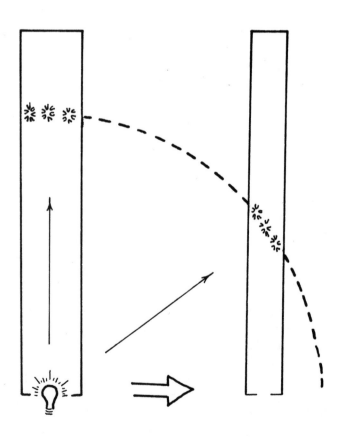

Figure 4–16. *The light clock must contract so that the tilted row of photons can reach across without stretching.*

Ganged clocks. The relationship between space and time would be easier to appreciate if somehow a clock could be devised that would fill all of space, rather than residing, as most clocks do, in one particular location. Lacking such a grand timepiece, you will have to be satisfied with clocks at different locations in space, ganged together so that all read the same time. How can you gang separate clocks to work synchronously?

Let the clocks be six-tooth gears. Each tooth represents an hour. (If you think a 6-hour clock odd, just wait until the metric maniacs compel you to live a 10-hour day.) The gears are mounted at equal intervals on a long stationary bar. The gears are given one-sixth turns periodically by pawls that are also located at equal intervals. The pawls are attached to another bar that slides smoothly and continuously to the right. The pawls are so spaced that they arrive at and turn all the clock gears in unison.

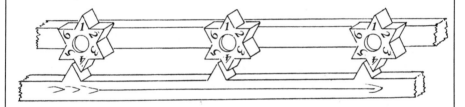

Figure 4–17.

In making this mechanism, how far apart would you space the pawls? Offhand you would answer, "The space between the pawls should equal the space between the clocks." But remember that the pawls and the bar to which they are attached are in motion, sliding to the right, and it has been established that material things (as well as the spaces between them) are measured as having shrunk in their direction of motion. Thus, if you want the space between the pawls to match the space between the clocks when the mechanism is running, that is, when the pawl bar is sliding, you had better make the pawl bar so that the space between the pawls is longer than the space between the clocks. That way when the bar shrinks due to its motion, the pawl spacing can come to be equal to the clock spacing.

Now consider how this mechanism would be measured by aliens, perhaps ants, riding along on the pawl bar. To the aliens the pawl bar is stationary. To the aliens the clock gears and the bar upon which they are

Figure 4–18.

mounted are in motion. The thing is moving to the left at the same speed you measure the pawl bar sliding to the right. But here comes the cute stuff.

Since the pawl bar is stationary to the aliens, the aliens measure the proper spacing between the pawls, which is larger than the proper spacing between the clocks. Moreover, the aliens measure the space between the clocks as shrunk, because they, along with their bar, are in motion relative to the aliens, sliding to the left. There is no way the aliens can measure all the pawls and all the clock gears ever coming into coincidence together. As measured by the aliens the clocks are therefore inexorably desynchronized. And these are the very same clocks I ganged together so that all would turn in unison. This example just recapitulates something you already know.

Another thing you already know also emerges. The aliens measure the moving clocks not only as desynchronized but as running slow. It is easy to see why. For a particular clock to pass from one o'clock to two o'clock, that clock must slide through a distance equal to the space between pawls. The aliens measure the clocks sliding between the pawls at the same speed that you measure the pawls sliding between the clocks. But the space between the pawls as measured by the aliens— the proper space—is longer than as you measure it. So it takes more time for the clock to get from pawl to pawl as measured by the aliens than as measured by you. This means that when a clock is perceived in motion, it is measured as taking more time to get from one o'clock to two o'clock than when the clock is perceived at rest.

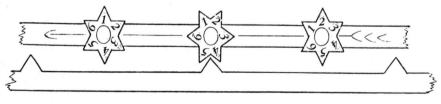

Figure 4–19.

Beating the limit. Suppose the speed of light is 100 miles per hour,* and a planet is moving through space at 99 mph. On that planet is an old car. If the car drives at just 2 mph, its total speed through space will be 101 mph. It will have beat the speed limit.

Why can't this happen? Does something hold the car back so that it can't go 2 mph? Nothing holds the car back. As far as aliens on that planet are concerned, they think they are at rest. Everything there goes as if they are at rest. The old car can certainly go 2 mph.

But who measures off the 2-mile stretch of highway? The alien does. And his planet is moving at 99 percent of the speed of light, so his 2 miles might seem to you to be 2 inches. And who times the hour? The alien does. But his clock is running slow. The alien hour might last a year.

So the alien thinks the car's speed has been increased by 2 miles per hour. You measure it to be increased by only 2 inches per year. Add and add and add, but you can never get to the speed of light.

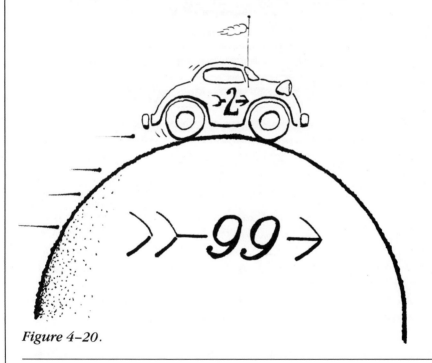

Figure 4–20.

*Rather than converting to the metric system (which is two centuries old), we should convert, if convert we must, to a system that makes the speed of light a number like 100 or 1 million.

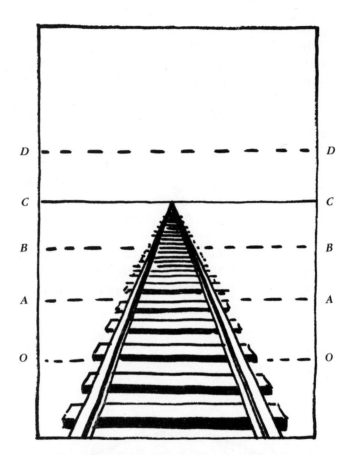

*Figure 4–21. The speed of light should be thought of as a **limit** with the exact meaning the word is given in first-year calculus. That means you can get as close to it as you desire, provided you don't touch it. If you think about situations that involve material objects touching the limit, you are asking for the same kind of trouble first-year calculus students fall into when they plug limiting values into their equations. To move through space at the speed of light, you must be light. There are even a few respected physicists who suspect light might not be at the exact limit. That is, even light might not travel at the "speed of light."*

It is like walking down an infinitely long railroad track. You start at line *OO,* you walk to *AA,* you can even walk to *BB.* In fact, nothing ever stops you. You can, if you want to, walk forever. But no matter how far you walk, you will never reach *DD.* So nothing in the world can go

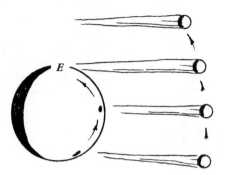

Figure 4–22.

faster than the speed of light? No. Material objects can't go faster, but many other things can be measured to go faster!

Yes, indeed, many things can go faster. For example, the moon's shadow when it runs off the edge of the earth at E is moving over the earth's surface infinitely fast. The spot of light made by a searchlight on the bottom of a sufficiently high cloud might exceed the speed of light if the searchlight is whipped around. The splash point where a wave hits a seawall can dash down the wall faster than the speed of light if the angle θ is small enough. If the angle is zero, the splash point will dash down the wall infinitely fast. And finally, the kink in the box hitting the conveyor belt, as perceived by the alien, runs along the bottom of the box faster than the speed of light.

Figure 4–23.

However, bear in mind that no material object is exceeding the speed of light. The particles of light (or darkness) that make up the spot (or shadow) are not the same particles from moment to moment. Likewise, the particles in the splash or in the bottom of the box do not move faster than light.

There is afoot an erroneous idea. It is that in physics the ultimate reality is a mathematical prescription, an equation. In fact, the ultimate reality is a little story or myth. To demonstrate this, I will provide a myth for the Special Theory of Relativity, which will eventually come to be taken as the reality.

Chapter 5

The Myth

Why?

In the foregoing chapters, I have laid out the logic of the Special Theory of Relativity as presented to the world by Einstein in his celebrated paper, published the year before the great San Francisco earthquake and fire. So the book could end here.

Nevertheless, some inquiring minds will persist in wondering: "Why?" "Why?" "Why?" "What makes time desynchronize?" "What makes space contract?" "What makes time run slow?" "What makes the speed of light unexceedable?" Though the inquiring mind will accept the standard explanation—which is that these spacetime transformations are required to make the speed of light constant—the inquiring mind will find no deep satisfaction in the explanation.

Continued persistence in inquiry will usually elicit this answer: "There is no way to explain what makes space contract," and so on. All that is known is: (1) the perceived speed of light is constant, and (2) Galileo's Dictum can't be beat. Those are empirical facts. Space contraction, and so on, follows as a logical necessity from the empirical facts. Moreover, there is no way at all to reason from your previous knowledge of space and time that weird things like space contraction will happen, for if there were, very powerful minds like Newton or Gauss would have deduced these weird things from their knowledge of the world. Case closed.

But the inquiring mind is yet unsatisfied. What does it want? I will tell you what it wants. The inquiring mind appreciates that direct deductions of answers to the "What makes...?" questions are impossible. WHAT THE INQUIRING MIND YEARNS FOR IS A LITTLE STORY OR MYTH THAT WILL GIVE THE FEELING OF BEING BACK HOME ON FAMILIAR, LOGICAL GROUND.

Good and Bad Myths

The invention of such a myth is not at all against the spirit of science — which has the job, first of all, of ENABLING THE INQUIRING MIND TO FEEL AT HOME IN A MYSTERIOUS UNIVERSE. However, not any old myth will suffice. There is a rule the myth must satisfy: The myth must SAVE THE PHENOMENON. That means that what is found in nature must be explained by the myth, and what is logically deduced from the myth must be found in nature. A good myth must also be EASY TO UNDERSTAND.

For example, the famous turtle myth is bad because it predicts that turtles are to be found at the South Pole; but when you go there you find penguins, not turtles!

On the other hand, the Quantum Field Theory myth is not good because virtually all talented physicists spend their energy advancing the subject rather than making it easy to understand. So psychic mediums fill the void, and many minds conclude that there is more reality in karma than in quanta.

Most areas of physics, which are thoroughly understood and regularly worked, are thought about,

Figure 5–1. The turtle myth is not bad because it seems stupid. It is bad because South Pole explorers don't find big turtles.

The myths of motion. The three laws of motion are associated with almost official myths. The first law, the inertia law, is taken to be the very essence of, if indeed not the ultimate cause of, sloth. The second law, the force-acceleration law, is taken as the paradigm of cause and effect. While the third law, the reaction law, is taken to be a natural expression of, perhaps the root cause of, vendetta.

Yet, some of these "official" myths are less than perfect. Take the second, for example. Is there a causation link between force and acceleration? A cause precedes its effect, but force and acceleration are not separated in time. Is force the cause and acceleration the effect? Consider the water hammer produced by abruptly closing a valve. Does the water's deceleration cause the hammer's force? Or is it the hammer's force that causes the water to decelerate? Or has the cause-and-effect myth beguiled you into tackling a chicken-or-the-egg question?

Figure 5–2. In the physicist's mind a force is synonymous with an arrow, but a force is not an arrow. That is just a good myth, and it is mildly shocking to awaken from it.

even by expert physicists, in terms of various good myths. For example, the force between static electric charges is thought about in terms of a sex myth: opposites attract, likes repel. Inertia is thought of as a natural, almost humanlike, lethargy of matter. Conservation laws and the Principle of Least Time or Least Action are cast as economics. The Second Law of Thermodynamics is thought of as the natural tendency for things to get messed up, Murphy's Law.

To understand the Special Theory of Relativity at the gut level, a good myth must be invented, and here it is.

The Story

Why can't you travel faster than light? **THE REASON YOU CAN'T GO FASTER THAN THE SPEED OF**

LIGHT IS THAT YOU CAN'T GO SLOWER. THERE IS ONLY ONE SPEED. EVERYTHING, INCLUDING YOU, IS ALWAYS MOVING AT THE SPEED OF LIGHT. *How can you be moving if you are at rest in a chair? You are moving through time.*

Why are the clocks moving through space perceived to run slower and slower as they travel faster and faster? Because a clock properly runs through time, not through space. If you compel it to run through space, it is able to do so only by diverting some of the speed it should use for traveling through time. As it travels through space faster and faster, it diverts more and more speed. How much can it possibly divert? The clock can divert ALL of its speed. Then it is going through space as fast as it possibly can, but there is nothing left for traveling through time. The clock stops ticking. It stops aging.

All this can be depicted by a diagram, which is essentially the cosmic speedometer diagram. Nothing can ever be done to alter the speed of anything. Only its direction of motion through spacetime* can be altered. At rest a thing is perceived to

Figure 5–3. This ornamental indicia from a Great Lakes Dredge and Dry Dock stock certificate illustrates Newton's myth applied to the moon's action upon the sea tides. One of the first and strongest objections to Newton's Theory of Gravitation was that it endowed inert matter with an animal characteristic: to be affected by and attracted to other distant matter through empty space without any physical connection, to flock together! The concept of inborn empathy, acting at a distance, seemed to be a regression to the mystic potencies, sympathies, and tendencies for which medieval science was ridiculed by Galileo. Newton did not like his own myth— but it worked.

---

*If you did not know the word area or if the word did not exist, you might coin the word spacespace to stand for area. Likewise, you might coin spacespacespace to stand for volume. Spacetime is coined to represent a surface that measures space in one direction and time in the other direction.

Figure 5–4. Moving through time. Locomotives usually depict motion through space. This sequence, illustrating the erection of a passenger locomotive, forcefully depicts motion through time.

speed through time, from O to A; but set in motion to the right, R, or the left, L, its velocity is tilted right or left. In the extreme, the velocity is tilted all the way to C, in which case all the perceived motion is through space and none through time. The diagram is easily calibrated. If the distance from O to A is 1 year, then the distance from O to C must be 1 light year—the distance light travels in 1 year.

Can you ever see your own proper clock stop ticking? No. You are always at rest with respect to yourself (unless you have out-of-body experiences), so you can never see your proper clock moving through space. Your clock, as you see it, always moves only through time, as it properly should. You can't even see your clock run slow, even though I can—when I see you traveling through space. Whichever way you are traveling, you take it to be the direction purely through time; and whichever way I travel, I take it to be the direction purely through time. But I don't always perceive you traveling purely through time, and likewise for your perception of me.

Have you wondered what it would feel like to travel at the speed of light? Well, tell me, how does it feel? You are doing it right now. It is that speed that is responsible for your feeling of time.

The question "Why can't things go faster than the speed of light?" can now be answered in another way, by asking another question:

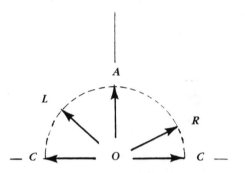

Figure 5-5. The thing is this: You are always traveling. And you are always doing so with a constant speed. Even when you stand still. When you stand still you are traveling through time, from O to A. If the speed is directed so as to carry you through space, like from O to R or from O to L, then the component remaining to carry you through time is diminished. If the speed is used entirely to carry you through space (at the speed of light), there is nothing left to carry you through time, and you move from O to C. Since your speed through spacetime is constant, the paths OA, OR, OL, and OC are of equal length and so the points A, R, L, and C form a semicircle around O. If you could go backward in time the semicircle would become a complete circle.

"Why can't time go by faster?" Have you ever been in a bad situation—being kept after school, or in jail, and wished you could somehow accelerate time? Well, if ever someone finds out how to make time go faster, to increase the speed of time, which means to increase the length of the cosmic speedometer's needle, that same discovery will also make it possible to go faster than light. And the converse is equally true.

Figure 5-6. If your speed through spacetime could be increased, that increase could be used to increase your speed through time or through space or a combination of the two. But no one knows how to increase it.

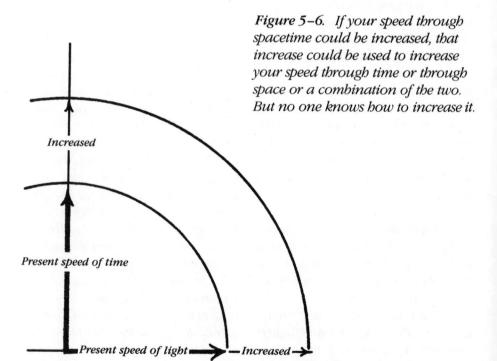

Increased

Present speed of time

Present speed of light ⟶ —Increased⟶

Through space and time. Several objects depart simultaneously from O. Each object has a different speed through space, and hence each has a different speed through proper time. But all have the same speed through spacetime, so all arrive on a circle in spacetime simultaneously.

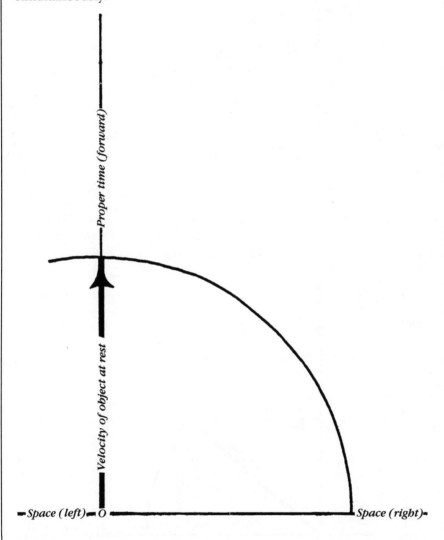

Figure 5–7. This object is stationary in space. It travels only through time. It ages one year while you also age one year. An object at rest in space has the maximum speed through time. It ages as fast as it can.

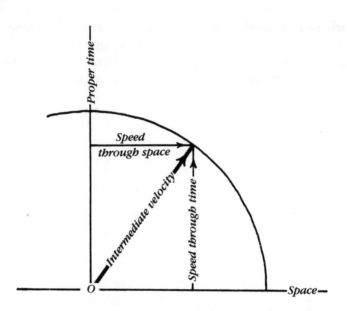

Figure 5-8. *This object is moving through space, so its speed through time is reduced. It travels through 0.6 of a light year of space and so ages 0.8 of a year of time. You age 1 year as you watch it happen.*

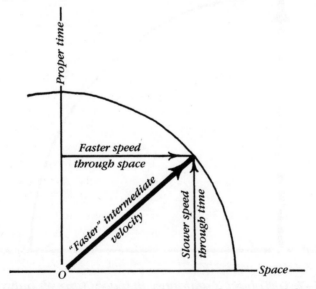

Figure 5-9. *As speed through space becomes faster, speed through time must become slower. This object travels through 0.8 of a light year of space while aging 0.6 of a year of time. You age 1 year while watching.*

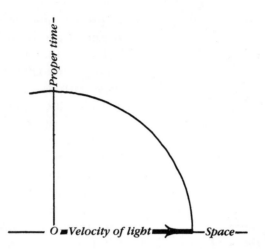

Figure 5–10. This object travels through 1 light year of space. You age 1 year as you watch. The object does not age at all. This object has the maximum speed through space, the speed of light. Its speed through time is zero. It is stationary in time. "Right now is forever," the photon said. This applies to gravitons and perhaps neutrinos, as well as photons. The numbers 0.6 and 0.8 were chosen for this example because $(0.6)^2 + (0.8)^2 = 1^2 = 1$. Any pair of numbers x and y such that $x^2 + y^2 = 1$ will work.

Out of Body

Let me now return to the previous comment on out-of-body experiences. If they were possible, you could see yourself moving through space. The closest you might actually come to such an experience is to watch an identical twin.

Peter and Danny are identical twins. Peter stays put at home. Danny goes out for a very high-speed run. Plotted on the diagram, Peter travels on path *p* straight through time. Danny travels on path *d*, which carries him on his run away from home and back again. Now Peter and Danny each have the same speed through spacetime (because all things have the same speed through spacetime—that's the myth). That means the distance Peter travels along path *p* always equals the distance Danny travels along *d*. The twins start together at the same place, as indeed all twins must. When Danny goes for his run, his path is diverted through space while Peter's continues directly through time. When Danny gets back, say, to *D*, Peter will be ahead of him at *P*. What does this mean? It means Peter is now ahead of Danny

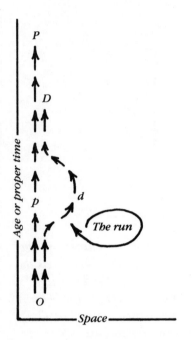

Figure 5–11. Peter and Danny each travel over the same length of spacetime, but Danny has more space and therefore less proper time in his journey than does Peter.

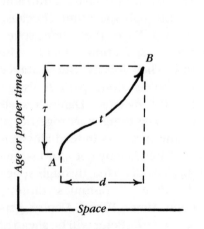

Figure 5–12. A and B are events separated in space and time.

in proper time or age. What does ahead in age mean? Astonishingly, it means that Peter is older than Danny, biologically older.* True, they started out as identical twins and, true, their speed through spacetime is forever equal, but Danny used some of his speed to carry him through space and so had less speed for travel through time. Thus, he fell behind Peter, who used all of his speed for time travel. There is no way Danny can ever catch up to Peter, unless Peter goes out for a run.

This makes a serious problem for people who claim out-of-body experiences. After being reunited with the body, does the age of the reunited corpus correspond to the age of the stay-at-home body or the age of the out-of-body traveler?

Two Kinds of Time

The diagram, which is a picture of the myth, dramatically illustrates that in relativity you must always be mindful of two aspects, or projections, of time: the time you observe to be required for a journey and the time the traveler reckons to be required.

The time you observe to be re-

---

*A detailed analysis of how this process is seen by the twins is given in the Special Relativity chapter of *Conceptual Physics* by P. Hewitt, Little Brown, Publishers, 34 Beacon Street, Boston, Massachusetts.

quired is equal to the length of the path on the diagram over which the traveler has passed, t, which is commonly called *coordinate time*. The time reckoned by the traveler is equal to how far up the diagram the traveler has moved, τ, which is commonly called the *proper time*. The proper time measures how much the traveler has aged. The coordinate time measures how much you have aged. The distance you observe the traveler has moved through space is equal to the sideways movement, d, which is commonly called the *space displacement*. If and only if the traveler is not traveling, that is, is stationary in space, then $t = \tau$ and $d = 0$. If and only if the traveler moves at the speed of light through space, then $t = d$ (if t is in years and d in light years) and $\tau = 0$; photons don't age because they travel at the speed of light.

This myth is not only conceptually appealing, but the picture can be measured, thereby telling precisely how much travelers age.

Question

Can you work this out? Peter and Danny are each 15 years old. On their birthday Danny takes off in his spacecraft, which moves at half the speed of light. He returns to Peter, who stayed at home, on the day Peter celebrates his 17th birthday. Biologically, Peter is 17 years old. Biologically, how old is Danny?

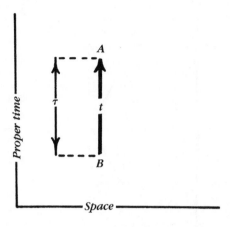

Figure 5–13. A *and* B *occur at the same place. They are separated only in time.*

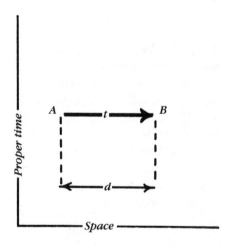

Figure 5–14. A *and* B *occur at the same* **proper** *time. They are separated in space and in* **coordinate** *time.*

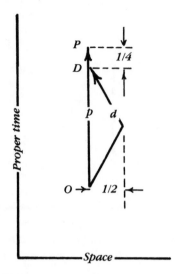

*Figure 5–15. If all motion is relative, how can you tell which twin really moved through space and should be younger at the reunion? The answer is that **not all motion is relative**. Only smooth, linear, unaccelerated motion is relative. And the round trip cannot be smooth, linear, and unaccelerated, because to come home the traveler must stop and reverse or must go in a circle. If you stop or turn, you can tell—even inside a sealed room.*

Answer

Draw a line 2 inches long straight through time; let it represent 2 years. That is Peter's path from *O* to *P*. Danny's path must also be 2 inches (2 years) long, but it is bent. Because he is moving at half the speed of light, Danny's path goes out into space ½ light year (½ inch) and back to *D*. Roughly, *D* is ¼ inch behind *P*. If 1 inch is 1 year, then ¼ inch is 3 months. That means Peter is 17 years of age, but Danny has 3 months to go. Biologically, Danny is only 16.75 years of age.

3D

The spacetime diagram used so far is overly restrictive. It shows only one direction in space (the right-hand direction). Also, and much more important, it shows only one space dimension. It is easy to draw a diagram with two space dimensions and it can be used exactly like the limited diagram. A thing at rest travels only through time from *O* to *A*. Light travels only through space (does not age), but it has its choice of directions, for example, from *O* to *B* or from *O* to *C*. And ordinary space travelers move on paths like *O* to *D*. All things have the same speed through spacetime, so all things departing simultaneously from *O* arrive on a hemisphere in spacetime simultaneously.

If the diagram were made complete, having three dimensions of space plus time, a four-dimensional picture would be required. While not impossible to imagine, it is not easy to draw. Incidentally, the idea of thinking of time as a dimension, like the space dimensions, originated long before Einstein.* The idea was popularized by H. G. Wells in his story "The Time Machine," which predates Einstein's work. Some nineteenth-century people even harbored the crazy idea that lost objects might be hiding in the fourth dimension. Anything plotted on a graph, like the Dow Jones average or the number of muggings per day, is represented as a dimension. But time has a special claim on being called a dimension. Why? Because time intervals change depending on how they are perceived, just as space intervals change depending on how they are perceived. The shadow of a yardstick is not always a yard long.

Now I am going to raise two seldom asked but salient questions. First, why did God make a world with three dimensions of space but only one dimension of time? Why not three time dimensions and only one space dimension? Or why not

*The intimate connection between space and time was more clearly recognized by our Roman and English ancestors—who divided the foot into 12 inches and the day into 12 hours—than by our metric contemporaries—who divide the centimeter into 10 millimeters, yet divide the day and year into 12 parts.

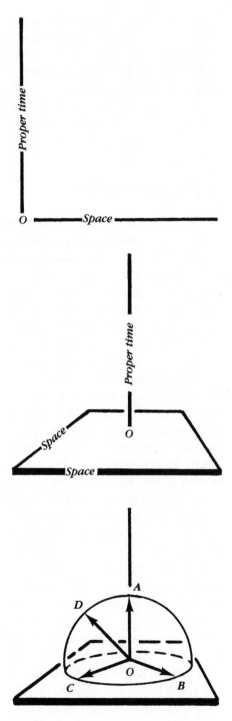

Figure 5–16.

an equipartition, with two space and two time dimensions? Second, why is it possible for you to move at will backward and forward (left and right) in the three space dimensions, while you are relentlessly and without choice drawn by the current of time in only one direction?

In the Beginning

The myth can be made to suggest answers to these questions. In the beginning God created four dimensions. They were all alike and indistinguishable from each other. And then God embedded atoms of energy (photons, leptons, etc.) in the four dimensions. By virtue of their energy, these atoms moved through the four dimensions at the speed of light, the only spacetime speed. Thus, as perceived by any one of these atoms, space contracted in, and only in, the direction of that particular atom's motion. As the atoms moved at the speed of light, space contracted so much in the direction of the atom's motion that the dimension in that direction vanished. That left only three dimensions of space — all perpendicular to the atom's direction of motion — and the ghost of the lost fourth dimension, which makes itself felt as the current of time. Now atoms moving in different directions cannot share the same directional flow of time. Each takes the particular current it perceives as the proper measure of time.

The world we mortals measure is only the shadow, the projection, of another world — the real four-dimensional world. Moreover, if you comprehend the shape and orientation of things in the real world, it is not difficult to figure out how their shadows will seem when cast upon our world.

Life can be blown into these words by means of a diagram. The first sketch shows the United States as you measure it whan at rest with respect to the country. The US is simply moving directly through time. Its projection on your space dimensions is not affected by the aging. The second sketch shows the US as measured by you if you are flying to the west (or the US is flying to the east) at about three-fourths of the speed of light. The speed of the US through spacetime is unaltered, but its direction of motion is now tipped eastward so that three-fourths of that speed is through space. The proper space dimensions of the US must be perpendicular to the direction of its motion through spacetime. But you don't measure the proper speed of time or proper space dimensions of a thing moving past you. You measure only the shadow of the proper speed of time and proper space dimensions as projected on your time and space dimensions. So you measure that the speed of aging for the US is reduced and you measure that the east-west dimension of the US has contracted, while the north-south dimension is unaltered. And as you

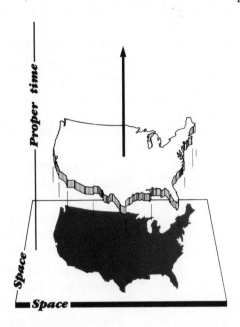

Figure 5-17. The United States as measured by an observer stationary with respect to the land. The United States moves directly through time, but its shadow, or projection in space, is not affected by its motion through time.

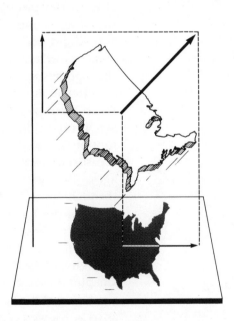

Figure 5-18. This picture completely summarizes the Special Theory of Relativity. It also provides a myth to account for the cause of the strange relativistic space and time effects.

watch, the whole country drifts eastward at three-quarters of the speed of light.

What Makes It Shrink?

But the myth has yet more to portend. As a literal and logical consequence of tipping the space dimensions so that they will be perpendicular to the direction of motion through spacetime, you put the western part of the US ahead of the eastern part in age. This is nothing but the previously discussed desynchronization of time.

Why does a yardstick shrink when set in motion? Just as a clock properly runs through time and is able to run through space only by diverting some of the speed it should properly use for aging, so also a yardstick properly reaches through space, not time. If you set the stick in motion, you couple it to reach through time, due to the desynchronization arising from the motion. The ends of the yardstick, as measured by you, are at different ages. The stick is able to reach through time only by diverting some of the length it should properly use for reaching through space.

Recall the description of the shrinking of the fleet of spacecraft in Chapter 3 (see page 41). It was as if a painting of the fleet was rotated so as to be viewed obliquely. The only problem was that the fleet wasn't rotated in space; its line of march through space was never changed.

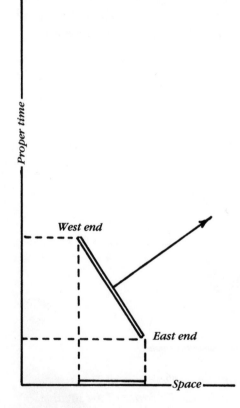

Figure 5-19.

Rotations require at least two dimensions. It is familiar for things to be rotated between two space dimensions, such as from the up-down dimension to the right-left dimension. But the space fleet was not rotated between two space dimensions. Rather, it was rotated from a space dimension to the time dimension. But rotated it was indeed!

Question

A super-starship spacecraft is 186,000 miles long and moving past you at nearly the speed of light. On board the craft all clocks are synchronized to the ship's time. As the ship flies past, you measure the ship's clocks at the nose end and the tail end to be reading two different times. What will the time difference be?

Answer

At rest, Ship 1 moves only through time. There are 186,000 miles between its ends, but both ends are at the same time. Moving at the speed of light, Ship 2 moves only through space. There are zero miles between its ends—both ends are at the same place (or almost the same place if it is almost at the speed of light). But there is now a 1-second time difference between the nose end and the tail end of the ship. One light second equals 186,000 miles.

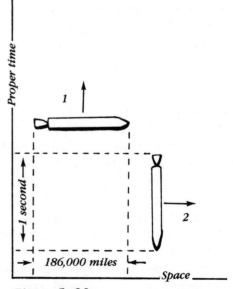

Figure 5–20.

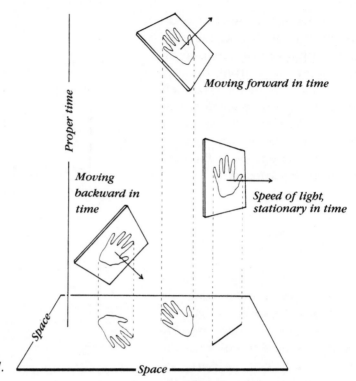

Figure 5–21.

Question

Suppose that one day you came upon a thing moving backward in time. Could you recognize that it was moving backward?

Answer

First of all, the thing moving backward in time regards itself as moving forward in time and you as moving backward. If the backward-moving thing has structure, so that it can identify proper left and right ends on itself, those ends are inverted when projected into your space. See Figure 5–21. A right hand traveling backward through time is perceived by you as a left hand. It is like looking at the hand in a mirror.

To complete this chapter, I am going to give you direct proof that the myth saves the phenomenon. Specifically, that a flash of light originating in the middle of a stick arrives at the ends of the stick at the same stick time, regardless of how the stick moves. If you dislike proofs, skip to the last section of this chapter, "A Parting Word."

The Myth Proven

Suppose the stick is 2 light years long and stationary in space. In 1 year the ends of the stick move from A to B and from G to H, each end going through 1 year of time, say 1 inch in Figure 5–22. During that year part of the flash that originated at O moves from O to A, and part of the flash moves from O to G, each going through 1 light year of space, also 1 inch in the figure. If the flash occurred on the first day of 1984, parts of the flash would arrive at the ends of the stick on the first day of 1985. However, note that the stick end is at B in 1985, while the photon of light from the flash is back at A in 1984. A and B cast the same shadow in space and have the same position in space. Both photon and stick have moved 1 inch from their initial positions, the photon from O to A and the stick from A to B, so both have passed through the same amount of coordinate time—the time measured by you. But, their proper times differ by 1 year. The photon did not age, but the stick did.

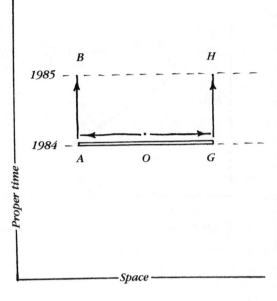

Figure 5–22.

Now suppose the same stick is moving through your space; it is always at rest in its own space. To save the phenomenon, it is required that light from a flash originating at O on the first day of 1984 arrive at the space positions occupied by each of the stick ends when those ends reach the first day of 1985, recorded according to the stick ends' proper time.

That means the flash occurs when the center of the stick passes the 1984 line (see Figure 5–23). The left end of the stick moves from A toward B. Part of the flash moves from O toward C, which is a point directly below B. IT IS NECESSARY TO PROVE THAT THE DISTANCE

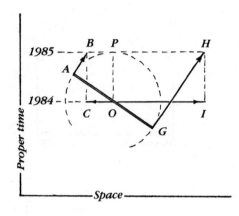

Figure 5–23.

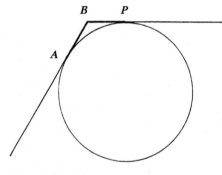

Figure 5–24.

FROM *A* TO *B* IS EQUAL TO THE DISTANCE FROM *O* TO *C,* that is, that the light arrives at the space position of the stick's left end when that end reaches the proper first day of 1985. LIKEWISE, IT IS REQUIRED THAT THE SAME CONDITIONS BE PROVEN AT THE STICK'S RIGHT END, namely, that the distance from *G* to *H* is equal to that from *O* to *I.* The point *I* is directly below *H.*

If *P* is directly above *O,* then the distance *P* to *B* is equal to the distance *O* to *C,* and likewise, *P* to *H* is equal to *O* to *I.* Also, the distance from *O* to *A* and *O* to *G* is 1 light year and the distance from *O* to *P* is 1 year, so all are of equal length. Thus, points *A, P,* and *G* lie on a circle and the lines *A* to *B, B* to *H,* and *H* to *G* are tangent to the circle (because $\overline{AB}$ is perpendicular to $\overline{AO}$, $\overline{BH}$ is perpendicular to $\overline{PO}$, and $\overline{HG}$ is perpendicular to $\overline{OG}$).

Now, if tangent lines are drawn from any point, like *B,* kissing the circle at *A* and *P,* then the distance *A* to *B* must be equal to the distance *P* to *B.* It then follows that *A* to *B* equals *O* to *C* (since $\overline{PB} = \overline{OC}$). Likewise, *G* to *H* equals *O* to *I* (since $\overline{PH} = \overline{OI}$). Therefore, what was required to be proven is proven. So the myth saves the phenomenon (MSP).

The diagram clarifies the essence of Einstein's "trick." Sure, it takes less of your (coordinate) time for the light to reach the left end of the stick than to reach the right end. But the aliens on the stick reckon by their proper time, and the stick was so tilted that it took less proper time for the left end to reach 1985 than for the right end to reach 1985. Desynchronization gave the left end a head start. Desynchronization was the key.

A Parting Word

You can measure miles in years and measure years in miles (astronomers call 6 million million miles a light year) because the speed of light establishes a constant conversion ratio between space and time. So I can rationally get away with talking about clocks running through space and yardsticks reaching through time. Nevertheless, you can't help but wonder — is The Myth for real?

Are you truly moving all the time, even when you sit still? Could it really be that you are not only perpetually moving, but also perpetually moving at the speed of light?

A similar question regarding the reality of the earth's motion was put to Nikolaus Copernicus. His 500-year-old answer will suffice for the present question.

Copernicus (or perhaps one of his assistants) wrote, in effect: "It is ridiculous to believe that the heavy earth is or could really be moving, and moving perpetually. But if you will just suppose it is so moving, it makes calculations easier to do and the world easier to comprehend."

So also let it be with The Myth for Special Relativity.

Question

A train moves past you so fast that you measure the engineer's clock running at half speed. The engineer measures your clock running at

a half speed.
b normal speed.
c double speed.

Answer

The answer is **a.** The situation must be completely symmetrical. If things are different for you or the engineer, that difference would tell who is **really** moving and that would be a violation of the Principle

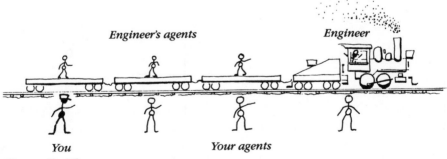

Engineer's agents *Engineer*

You *Your agents*

Figure 5–25.

of Relativity. But how can it be that each party perceives the other clock to be slow? After all, if my clock is slow relative to yours, then yours is fast relative to mine.

To understand the situation, you must appreciate that you don't watch the engineer's clock; your

local agent watches it. And the engineer doesn't watch your clock; his local agent does. (Remember the difference between SEEING and MEASURING? See page 54.) The engineer's agents have synchronized watches as measured by him, but not as measured by you. Your agents

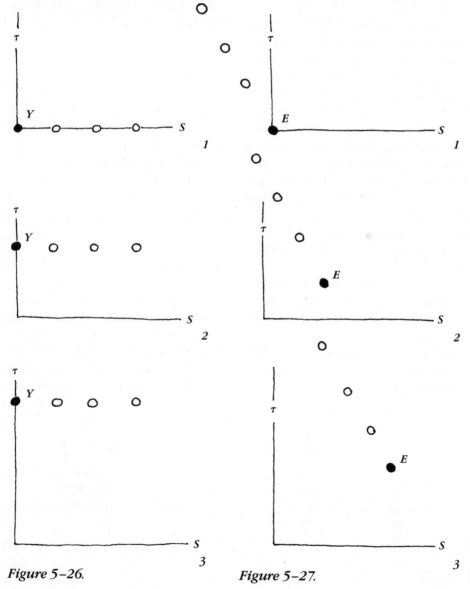

Figure 5–26.

Figure 5–27.

have synchronized their watches as measured by you, but not as measured by the engineer.

Figure 5–25 shows the train before it gets up speed and shrinks. Figure 5–26 is a sequence showing you and your agents passing only through time. Figure 5–27 is a sequence showing the engineer and his agents in high-speed motion. Note the desynchronization—the agents' watches are ahead of the engineer's and all are running at half of normal speed.

Figures 5–28 to 5–30 show the motion of two lines. One line represents you and your string of agents, the other represents the engineer and his string of agents. Both lines move through spacetime at the same speed.

As the locomotive passes you, Figure 5–28, both you and the engineer read 1 January 1984 on your own respective pocket watches. In the time between Figures 5–28 and 5–29, your watch advances ½ year, but the time reading on the watch of the engineer's local agent, E_1, who is passing you, is 1 January 1985. The local agent reports to his boss, the engineer, that your watch is 6 months slow.

In the time between Figures 5–29 and 5–30, your watch and the watches of all your agents advance another ½ year, so all read 1 January 1985. Yet your local agent, Y_1, who is next to the locomotive, reports to you that the engineer's watch is 6 months behind.

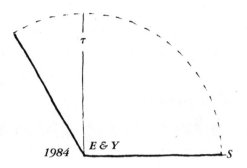

Figure 5–28.

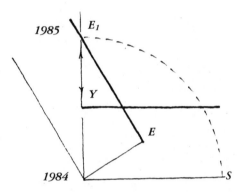

Figure 5–29.

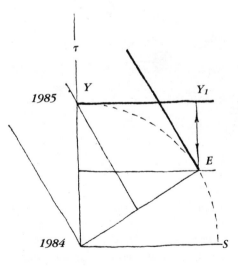

Figure 5–30.

Chapter 6
The Big Bang

Chapter 6 is not essential to understanding relativity and can be skipped. The chapter shows the application of relativity theory to the system of the world, specifically the Big Bang cosmology. It is included here because, having read this far, you now have an adequate background to understand it and because I suspect most people interested in relativity are also interested in astronomy, Moreover, it is always instructive to see theory applied. As Newton said, "In learning the sciences, examples are of more use than precepts."

Genesis

The universe began with a Big Bang. Presently, that is the most in vogue, widely known, and widely believed (by scientists and laymen alike) story of Genesis. The idea is outlined in every astronomy text published during the last half century. It is based on the observed fact that all the distant galaxies seem to

Figure 6–1. The Big Bang illustrated, as described by hundreds of planetarium and television shows and by thousands of science teachers and books.

be flying apart. As additional evidence, the texts published in the last decade cite high-frequency radio wave (microwave) radiation coming from space, which was first detected at the Bell Telephone Laboratories. This radiation is supposed to be the remnant of the flash emitted by the big explosion, or emitted very shortly thereafter by the fireball that followed the bang.

Aside from the truth or fiction of the Big Bang Theory itself, you should rightly ask how Bell Labs can still detect any part of the Big Flash. After all, the photons from the flash must fly out at the speed of light. The massive galaxies also fly out, but at various speeds less than the speed of light. By this late date the flash photons should be way beyond the galaxies, and Bell Labs is *in* one of the galaxies. So how can Bell detect any Big Bang photons? If you were riding on a little fragment of shrapnel flying out from the explosion of a hand grenade, you would not expect to hear the bang long after the grenade exploded. This is not the only dilemma raised by the Big Bang story.

The observed speeds and distances of the expanding galaxies make it possible to figure out approximately how long the expansion has been in progress, that is, how long ago the Big Bang happened. And that immediately makes it possible to figure out the maximum size of the universe. The universe can be no bigger than its age multiplied by the speed of light.

That means the universe has an outer edge or outer surface someplace. So you could ask, "What is beyond the outer boundary?" Yet, if you read further in the general astronomy texts, you come to something called the "Cosmological Principle," which states that all parts of the universe are the same, or at least are the same when they are of the same age. So there can be no special place like the edge.

If you turn on your imagination you might suppose you could get out of these difficulties by invoking some curved space hocus-pocus. But reading yet further, you find that the astronomers are undecided about what kind of curvature space should have. Should it have positive curvature or negative curvature? Lacking any definitive facts from the astronomers, let's choose the middle ground, zero curvature. It is the most conservative bet and the easiest to picture, since zero curvature means flat space.

I will now show how both the fireball flash and the edge-of-the-universe paradox can be cured by directly applying simple, relativistic effects in plain, flat space.

Looking Back

First, I will show how it is possible, even still, to detect photons that originated in the Big Bang fireball. For a short time after the Big Bang explosion, all the matter in the universe must have been glowing

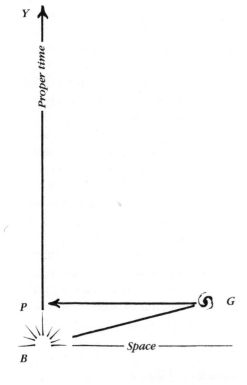

Figure 6–2. You are always older than the galaxies you see. Age refers to the age of the matter in you and in the galaxies, as counted from the Big Bang. In this sketch, the distance from P to Y measures how much older you are than the galaxy at G.

white hot and therefore emitting photons. As you look at ever more distant objects in space, you also look back ever further in time. Can you look back far enough to see the Big Bang fireball?

Offhand, the answer seems to be, "No, you can't see that far back." Offhand, suppose the universe is thirty billion years old, and suppose a fragment is moving away from you as fast as it can, that is, at the speed of light or almost at that speed. Further suppose that fifteen billion years after the Big Bang, that fragment emits a photon back toward earth. The photon will require another fifteen billion years for the return trip. This line of thought indicates you can only see halfway back to the Bang. Fifteen billion years after the Bang the fireball would be long gone. So how can Bell Labs detect anything?

Bell Labs can detect Big Bang photons because one vital consequence of relativity has been omitted from the foregoing offhand analysis. The fragment flying away from you at nearly the speed of light does not age fifteen billion years. That is, as you perceive it, it does not age fifteen billion years. In fact, if it is flying away at exactly the speed of light, it would not age at all, relative to you. Since it has not aged, you see the fragment as it was at the moment of Genesis. If the fragment is flying away from you at a speed slightly less than the speed of light, it has aged only slightly, so you see it as it was a few hours or years after the Big Bang, that is, during the era of the fireball.

All this can be easily illustrated on a spacetime diagram. You travel straight through time from the Big Bang at B to your present age at Y. The distant galaxy also starts at the Bang and flies away from you along line BG. At G it emits a photon that returns to you at P. The distance from B to Y equals the distance from B through G to P. That distance measures the lapsed coordinate time since the Big Bang, so the photon gets back to P exactly when you get to Y. But your age is equal to the length BY. The photon's age, as well as the distant galaxy's age when the photon was emitted, is equal to the much shorter length BP. So you see a very young galaxy. This situation is almost a replay of the traveling twin story.

If you look back far enough to see the fireball, you will be looking at a time before the galaxies formed. So G should be a fragment of the fireball, not a galaxy. As the universe ages, you must look at fragments moving ever closer to the speed of light in order to go back far enough to see the fireball era.

The fireball fragment you look at now must be moving away from you at almost the speed of light. So the radiation you detect from it is enormously redshifted. If you redshift light enough it becomes infrared (heat radiation), and if you redshift it yet more it becomes radio waves. That's why the Big Bang flash is heard rather than seen. In the ex-

treme case of fragments moving away from you at the speed of light, the emitted radiation would be redshifted to infinitely long waves and so be undetectable.

Age Is Impossible

If you will reflect on the idea of the expanding universe, it will force you to change one of your basic preconceptions of time, specifically the concept of age. What is the age of the universe now?

You can't really answer that question. You can say what the age of this galaxy is now, but what about the present age of a galaxy seen to be in rapid motion relative to yours? You can't assign an age to the whole universe if its parts are in motion relative to each other. Each part has its own proper age.

Each part perceives itself as older and the other parts as younger, and ever younger as they are ever remoter. Note that this is **not** because it takes time for the light to get from the distant parts to you. It is because the ever remoter parts are in ever more rapid motion relative to you.

This relative time effect applies to any mechanism that has moving parts: The faster a part moves the slower it ages. In principle it is impossible to assign one age to a mechanism with moving parts. For most mechanisms this is unimportant practically, but for the mechanism of the universe as a whole it is of paramount importance.

Figure 6–3.

The Ultimate Horizon

Relativity not only has vital consequences for the measured age of rapidly moving galaxies, but also for the space you measure between them. Suppose the galaxies spaced themselves in a line (a one-dimensional universe) and set themselves in motion in this way: Galaxy X locates itself 1 mile from Galaxy Y and is moving away from Y at 1 mile per hour. Galaxy W locates itself 1 mile from Galaxy X and is moving away from X at 1 mile per hour. And so on. How would you measure this universe from the standpoint of Galaxy Y? While the pair of galaxies W and V perceive themselves to be 1 mile apart, you would not perceive this proper separation. Since the pair is in motion relative to you, your perception of the pair's separation would be less than 1 mile.

The string of galaxies might be infinitely long, each moving a mile per hour faster than its neighbor, but you would not measure the infinitely distant galaxies moving infinitely fast. The most remote galaxy could be measured moving no faster than the speed of light relative to you. And the measured separation between the pair moving at the speed of light would shrink to zero.

All adjacent galaxies find themselves equally spaced and moving with the same speed relative to their adjacent neighbors. This seems to satisfy the Cosmological Principle. Might this be a model for the expanding universe? Not quite.

In an expanding universe the proper separation between adjacent pairs of galaxies increases with age. If the proper separation between all pairs is at once equal, that means that the whole universe comes to be the same age simultaneously; and that is impossible. However, with the aid of the spacetime diagram you can draw a model that satisfies all requirements.

The Big Bang is at B. The galaxies

Figure 6–4.

fly out along the lines *BY, BX, BW,* and so forth. When each galaxy arrives at the horizontal line drawn through *Y,* each one has an age equal to the proper time lapse *BY,* so at that age, the galaxies are properly separated from their adjacent neighbors by 1 mile. But be careful; they don't all reach that age simultaneously. The time you measure to be required for your galaxy to reach the mile-apart age is equal to the length *BY.* The time you measure to be required for Galaxy *V* to reach the mile-apart age (the coordinate time) is equal to the length *BV.*

In this model there must be an infinity of galaxies, so no galaxy can be the last one at the edge of the universe, just as no number can be the last number. However, if you were to survey the universe at the time your galaxy reached *Y,* you would find the other galaxies in a space-time circle of radius *BY* centered on the Big Bang. Each galaxy would have a different proper age and distance from you. Near *A′* their proper age would be nearly zero, the age of the fireball. At *A′* you would perceive the edge of the universe. How can the universe be perceived to have an edge when (in this model) the universe extends on forever? The same way an artist sees the edge of a flat plane at the horizon when the plane extends on forever.

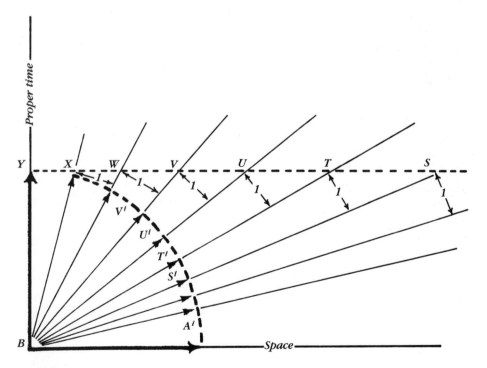

Figure 6-5.

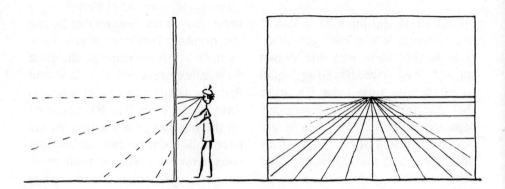

Figure 6–6.

**Space makes geometry; space
and time make animation; mass,
space, and time make the universe.**

Chapter 7
The Third Leg

Mass

At the conclusion to Chapter 2 it
was emphasized that jacking around
with the conventional ideas of space
and time means jacking around with
the foundation of all physics. Having
jacked around, let's face the consequences.

It is a remarkable fact that all
physical quantities (force, momentum, energy, speed, torque, etc.) can
be expressed in terms of only three
physical measures: a measure of
space (like the meter), a measure of
time (like the second), and a measure of mass (like the kilogram). This
fact is celebrated in the seal of the
American Institute of Physics by the
meter stick, the seconds pendulum,
and the kilogram mass. Having altered space and time, attention
must now go to the third leg of the
tripod on which physics sits—the
mass leg.

Mass makes its mark on your con-
sciousness in two ways. First by its
weight, which has to do with gravity, and which changes from place to

Figure 7–1.

Figure 7–2. Comets are most easily seen just after sunset or before sunrise. Their tails extend away from the sun and into the darker part of the sky.

place (an increase on Jupiter, a decrease on the moon). Gravity will be discussed in chapters 9 through 12. Second, mass makes its mark by inertia, which means how hard it is to start or stop its motion (disregarding all friction). Even in distant space where there is little gravity and zero friction, objects have their inertia; that's why spacecraft require rocket engines for midcourse maneuvers.

Momentum

How hard it is to start or stop something depends jointly on how fast you want it to go (or how fast it is going if you want it to stop) and how massive the thing is. For example, the punch in a fist depends jointly on the speed and the mass of the fist. The old Bourbon Street bouncer in New Orleans, who can throw a punch only half as fast as he could in younger years, still puts the original wallop in his punch by putting a roll of quarters in his first so as to double its mass. Physicists call punch momentum, so

$$momentum = punch$$
$$= mass \times velocity.$$

If you get hit by something, and that something has punch, then it must also have mass. This statement does not seem overly profound, but it leads immediately to an insight.

For five centuries it has been known that a comet's tail always points away from the sun. When a comet moves away from the sun it moves tail first. The age-old explanation has been that sunlight exerts a force (called radiation pressure—the same force that prevents the sun from collapsing under its own gravity) on the dust and ice crystals in the comet's tail. Radiation pressure literally pushes small particles away from the sun. If sunlight exerts a force on the tail, then sunlight must carry punch. If light can punch, it must have mass! And the brighter the light, the heftier the mass. Note that all kinds of waves—sound waves, earthquake waves, and light waves—carry energy. But light waves alone carry mass and momentum. Air and earth have mass, but that is not the wave's mass. Air and earth masses are there even when waves are not. Moreover, sound and earthquake waves do not carry momentum, but rather exert equal amounts of positive and negative force, so that their total effect amounts to zero net force and zero net momentum. Sound will not blow smoke through the air and earthquakes are shakers not bulldozers.

In focusing your concept of what mass is, you have shed new light on what light is. In fact, you can even carry this line of thought a step further. If light energy has mass and light energy can be transformed to another form of energy, such as heat or electric energy, then that other form of energy must also have mass. Otherwise, mass would vanish. Since all forms of energy can be trans-

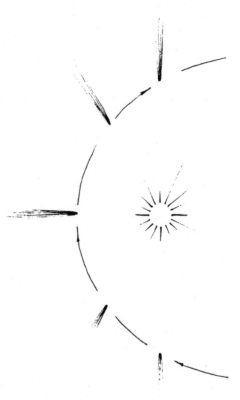

Figure 7–3. Some big comets have heads larger than the sun and tails longer than the radius of the earth's orbit. The tail grows as the comet nears the sun.

formed one to another, you could argue that all energy must have mass! I will return to this idea in the next chapter.*

Punch Out

Thus far the discussion of mass (and light) has been based completely on ideas established centuries before Einstein. But now I am going to tell a tale illustrating how relativity creeps into the act. Suppose two trains (traveling at almost the speed of light) rush past each other in opposite directions. On board one train is one of the identical twins, Peter, and on board the other train is the other twin, Danny. As occasionally happens, the twins are having a little fight. Peter intends

Figure 7–4.

---

*It might now be argued that sound and earthquake waves must carry some mass, since they carry energy. And that is correct. But, and this is even more important, it is a hindsight.

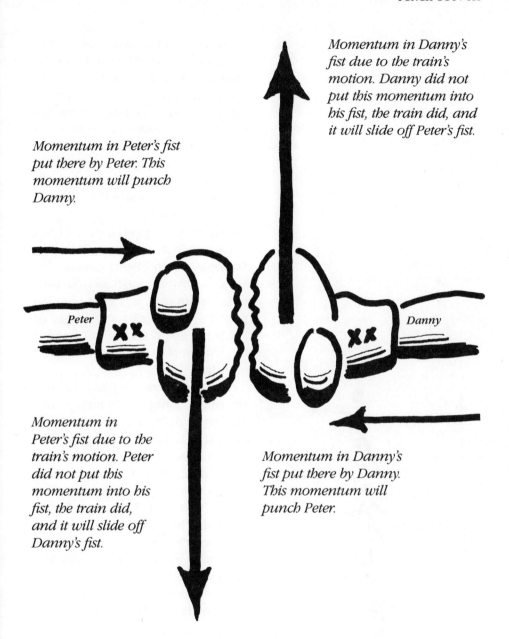

Momentum in Danny's fist due to the train's motion. Danny did not put this momentum into his fist, the train did, and it will slide off Peter's fist.

Momentum in Peter's fist put there by Peter. This momentum will punch Danny.

Peter

Danny

Momentum in Peter's fist due to the train's motion. Peter did not put this momentum into his fist, the train did, and it will slide off Danny's fist.

Momentum in Danny's fist put there by Danny. This momentum will punch Peter.

Figure 7–5.

to haul off and punch Danny when he passes him. And Danny has the same intention for Peter.

It takes some time to get a punch thrown from your body to the full extension of your arm. Let's say it takes $\frac{1}{10}$ of a second. Peter perceives Danny's punch coming. But remember, Danny is on a train rushing by Peter at nearly the speed of light, so everything happening on Danny's train is perceived by Peter to happen in slow motion, including Danny's punch. Peter thinks that Danny's slow-motion punch has little momentum in it, while his own fast punch has lots of momentum. So Peter thinks his fist will hit Danny's fist and push it back into Danny's mouth.

However, Danny's perspective is the mirror image of Peter's. Danny perceives time to be going slowly on Peter's train, so Danny perceives Peter's $\frac{1}{10}$-of-a-second punch to require 2 seconds. Danny perceives his own punch in its proper time, $\frac{1}{10}$ of a second. Therefore, Danny thinks his fist has more momentum than Peter's, and expects that it will hit Peter's fist and push it back into Peter's mouth.

When the moment of truth arrives and fist hits fist, whose expectations are correct, Peter's or Danny's? The situation is completely symmetrical. Symmetry rules. They both can't get their own fists shoved back into their mouths, so neither does. It is a standoff; the fists come together with equal, but opposite, momentum.

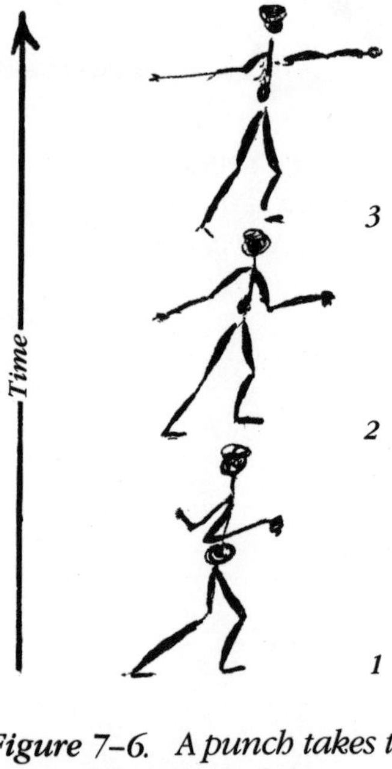

Figure 7–6. A punch takes time to throw. If the speed of time is reduced, the punch becomes a slow-motion punch.

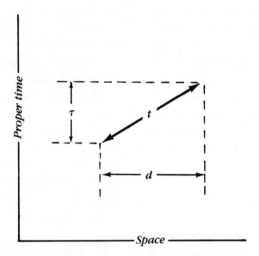

Figure 7–7. *While an object moves a distance,* d, *a clock attached to the object measures a proper time interval,* τ, *and your clock, not moving with the object, measures a coordinate time interval,* t.

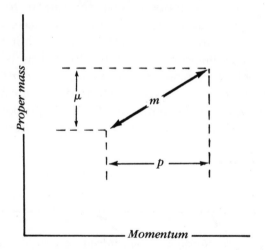

Figure 7–8. *The momentum of the object moving in Figure 7–7 is* p. *The rest mass of the object is* μ, *and you, not moving with the object, measure its moving mass—frequently called its dynamic or coordinate mass—to be* m. *Figures 7–7 and 7–8 are geometrically similar or proportional figures. In algebra that means:* d/p = τ/μ = t/m = *some proportionality constant,* κ. *As time goes by,* d, τ, *and* t *grow, causing Figure 7–7 to scale up. If the speed of the moving mass does not change, then* p, μ, *and* m *do not change, so Figure 7–8 does not change. Consequently, the constant of proportionality,* κ, *must grow as time goes by.*

Peter is amazed. How could Danny's slow-motion fist have as much punch in it as his? Peter reasons that the only way Danny's slow punch could have as much momentum in it as his fast punch is if Danny put a roll of quarters in his fist. Likewise, Danny reasons that Peter had a roll of quarters in his fist.

Relative Mass

In fact, neither twin had put quarters in his fist. Yet the fists acted as if there were quarters in them. This is a new relativistic effect. Just as motion desynchronizes and slows down time and just as motion shrinks space, so also motion affects mass. It causes the measured mass of a thing to increase as its speed increases. How much does mass increase? From the story you can see that mass must increase by the same factor that the speed of time (and speed of punches) decreases. That is, the longer the time you measure to be required for the punch to be thrown, the greater will be the measured mass of the fist.

This idea is easy to picture. The twin perceives himself throwing a punch in the proper time, τ. While he is throwing the punch you perceive the train moving a distance, d, and you measure a time, t, elapsing. And t is longer than τ; that's why the punch looks like slow motion to you.

Now the mass of the fist that you measure must be larger than the proper mass of the fist in precisely the ratio of t to τ. So draw a new diagram similar to the spacetime diagram and make the mass of the fist when it is not moving proportional to τ. This is called the *rest mass*, or *proper mass*, μ. Then the measured dynamic mass of the moving fist, m, is proportional to t. So t becomes m if τ becomes μ.

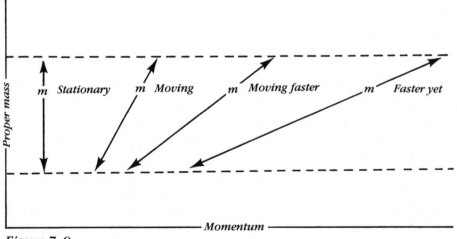

Figure 7–9.

It is easy to see that the dynamic mass (and momentum) of a thing becomes infinitely large as its speed approaches the speed of light, even though its proper rest mass remains unchanged. This explanation is sometimes given as the reason why things can't go faster than the speed of light—they get so massive that further acceleration becomes impossible. But this is not a complete argument. Why? Because, for example, the automobile engine that accelerates the auto rides with the auto and, so far as the engine is concerned, the car's mass remains the unchanged proper rest mass of the automobile. Moreover, some forces, like gravity, increase with increasing mass, so for them mass cannot limit acceleration. I will return to this issue shortly.

Relative momentum. When τ, *proper time,* becomes μ, *proper mass,* and t, *coordinate time, your time,* becomes m, *dynamic mass, as measured by you,* what becomes of d, *distance moved by the train?*

Well, d was the shadow of t, so it becomes p, the shadow of m. The length of the shadow of m depends jointly on two things: first, the length of m and, second, the ratio of the shadow's length to the length of m. That ratio is the same as the ratio of d to t (distance moved by train to time required for that motion), which is the train's velocity. So, p, the shadow of m, is equal to the fist's dynamic mass multiplied by the train's velocity. Mass multiplied by velocity is punch or momentum. That means p must be the momentum of the moving mass. Summing up, if proper time becomes proper mass, then coordinate time becomes dynamic mass and distance becomes momentum. Note that momentum p is not the punch directed by one twin against the other; it is the momentum in the fist due to the train's speed. That larger speed and momentum dwarf the smaller speed and momentum the twins are able to put into their fists.

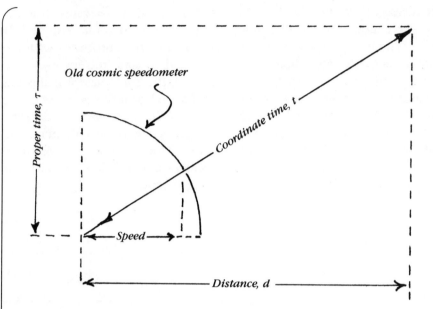

Figure 7–10.

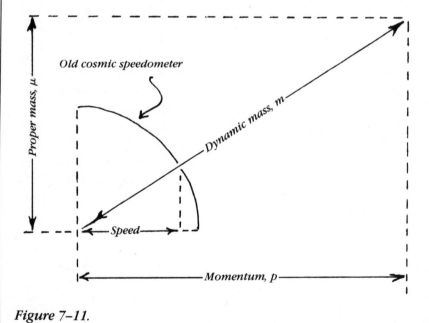

Figure 7–11.

Energy Has Weight

From the discussion so far, you might surmise that the measured dynamic mass of a thing somehow depends on its momentum or speed, but that is not so. Suppose you have a completely sealed box. Inside the box two lumps of clay move toward a head-on collision. After the collision they are stationary. Does the mass inside the box decrease after the lumps of clay stop each other's motion? How can any mass escape? The box is sealed. Or suppose you have a perfectly sealed box with a coiled spring and flywheel inside. You let the spring uncoil, which sets the flywheel spinning. The moving wheel must have more mass than the stationary wheel. But where did the mass come from? How did it get into the sealed box?

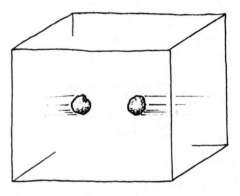

Figure 7–12.

In both cases no mass can leave or enter the sealed box. The mass in the box stays in the box. True, the stationary lumps of clay have less mass than the moving clay, but the "lost" mass is in that box "hiding" someplace. Where can it hide? What is in the box after the collision that is not in there before the collision? The heat produced by the impact. Perhaps the mass is in the heat!

The spinning flywheel certainly has more mass than the stationary wheel, but where inside the box can the mass have come from? What changed in the box when the wheel began to spin? The spring unwound. The mass must have come out of the spring! But no mass came out of the

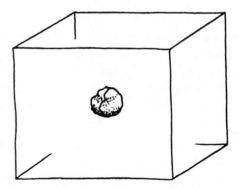

Figure 7–13. The collision does not change the total mass or total energy inside the sealed box.

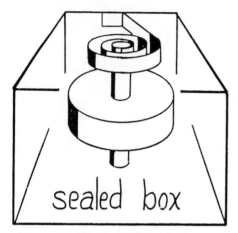

Figure 7–14. In this sketch the spring is wound tight and the flywheel is not spinning.

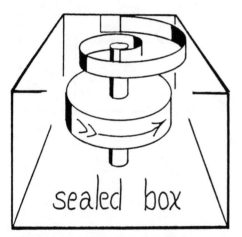

Figure 7–15. In this sketch the spring has unwound and the flywheel is spinning. The spring's loss is the flywheel's gain.

spring, only energy came out as it unwound. Then energy must have mass!

In both cases the energy—all the energy—remains inside the box. The lumps of clay gain in heat energy exactly what they lose in kinetic energy, so if the mass is in the energy, there is no change in the total mass inside the box. The flywheel gains its kinetic energy and mass from the spring, which literally means that a wound-up spring weighs more than an unwound spring!

Winding a spring adds no speed or momentum to it. Winding only adds energy. So you must appreciate that it is the addition of energy, rather than speed or momentum, that is responsible for the added mass. Added speed usually means added energy, but added energy does not always mean added speed.

This mass effect is a direct but unforeseen (it surprised Einstein) consequence of jacking around with space and time.

Well, what is the weight of heat? By how much does the weight of a pocket watch increase when you wind it up? How much does energy weigh? That's the subject of the next chapter.

Impedo. In the days of Galileo the distinction between energy and momentum was unrecognized. A body in motion was simply endowed with a substance called "impedo" (as an electrified body was endowed with "vitreous" or "resinous humor" — still called juice by electricians, and a heated body was endowed with "caloric").

When you throw a stone you can feel the impedo going down your arm from your muscle, through the pressure in the palm of your hand and into the stone. Once in the stone the impedo makes it fly. When the stone hits a wall some of the impedo goes into the wall, causing damage, and some goes into the air, causing sound. The stone, having lost its impedo, just drops to the ground. You can also feel the flow of impedo when you lift a heavy stone. But a serious argument developed about how much impedo was released by dropping the stone.

Galileo showed that if the distance a stone fell was increased four times, its speed at impact only doubled. (Galileo worked this out theoretically and it surprised him so much he was driven to test it experimentally. Contrary to popular belief, Galileo, like many theoretical physicists, preferred to think out, rather than do, experiments.) The French philosophers took this to imply that the impedo in the stone increased four times when its speed doubled. The English insisted that the impedo only doubled when the stone's speed doubled.

The argument was resolved by introducing the ideas of energy and momentum. Energy was French impedo; it increased four times while English impedo only doubled. English impedo was momentum.* But how could one substance, impedo, be two things: energy and momentum? It was exacty like the wave-particle duality question for light. The question was resolved by striking the word impedo from the language of physics, exactly as the word æther was struck from the language.

But the ghost of impedo is mighty (as is the ghost of æther). Its resurrection was given the cumbersome name "four-vector momentum," a vector that has three space components equal to the 3D momentum of a particle, and a fourth component through time equal to the energy of the particle.

---

*English thought was dominated by Newton, who based his mechanics on momentum; he called it *quantity of motion.* French thought was dominated by Lagrange, who based his mechanics on energy; he called it *vis viva.* Momentum being a vector lends itself to graphic illustrations. Newton's book has a diagram on almost every page. Energy is a scalar quantity, so Lagrange could get away without diagrams and he was boastful that his book had no pictures and lots of equations.

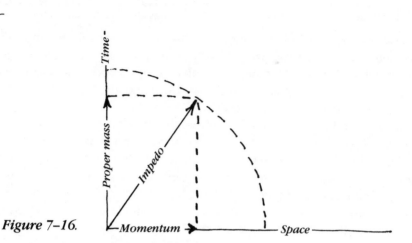

Figure 7–16.

The mass-momentum diagram of a wind-up toy or anything cut off from outside input allows the following interpretation. The impedo is a line of fixed length pointed in the direction of the object's motion through spacetime. All you can ever see are its shadows. Its shadow on space is called momentum. Its shadow on time is proper mass, which is proportional to proper energy. When rest mass is transformed to radiation, or vice versa, the impedo vector simply flips 90 degrees.

Incidentally, the concept of impedo can also be preserved in classical physics. It can be pictured as a slice of bread. The white is the energy, the crust is momentum, and the thickness of the slice is the mass. If the thickness does not change but the amount of white increases four times, the amount of crust doubles. If the cross-section does not change but the thickness doubles, both white and crust double.

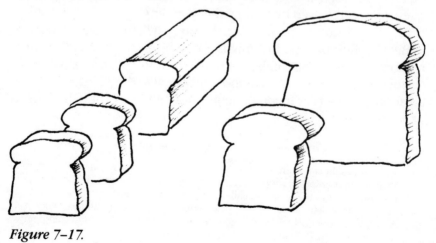

Figure 7–17.

Chapter 8
$E = mc^2$

Equivalent Energy, Equivalent Weight

How much does energy weigh? To answer this question, note first of all that a certain amount of energy, say a foot-pound* of energy, must have the same mass regardless of what kind or form of energy it is. It might be nuclear energy in a reactor or chemical energy in a battery or the kinetic energy in a spinning wheel or heat energy in a steam dome. It is not hard to see why this is.

Seal a nuclear reactor and a storage battery inside a box. Nothing can enter or leave the box. Now let the reactor put out a foot-pound of electric energy and let that foot-

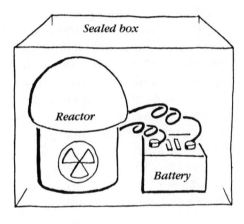

Figure 8–1. *If someone had a sensitive enough scale, you could measure how much energy was in a battery by weighing it uncharged and again charged. Such a sensitive scale does not exist presently.*

---

*A foot-pound of energy is the energy required to lift a pound weight 1 foot. But don't suppose the energy weighs 1 pound, because a foot-pound of energy can also lift 2 pounds ½ foot or ½ pound 2 feet. Other units of energy are: calories, ergs, joules, kilowatt hours, Btu's, electron volts, and, as I will show, ounces.

pound of energy be put into the battery. As the reactor puts out energy it must lose mass. But no mass can escape the sealed box. So where could the reactor's lost mass be? Only in the charged battery. The mass goes with the foot-pound of energy. Later, if something else, such as a light, receives the foot-pound of energy from the battery, it also receives the mass that goes with it. That means if the energy goes into light, so also the mass goes into light.

If the mass of 1 foot-pound of any one form of energy could be found, even under some very special and unusual circumstance, then the mass of 1 foot-pound of all forms of energy would be immediately known, because all foot-pounds of energy must have the same mass, regardless of form.

Masserators

Ironically, it is the mass of the energy given to particles, like electrons, traveling at almost the speed of light inside the 2-mile-long Stanford University linear accelerator (accelerators were once known as atom-smashers) that is easiest to calculate. The reason it is easiest to calculate the mass of the energy put into particles traveling at or very nearly at the speed of light is this: Added energy usually increases both the speed and mass of the thing receiving the energy. But if the thing is already going as fast as it possibly can (or almost as fast), the added energy can hardly increase its speed any more. Only the mass can change.

The Stanford accelerator is sometimes called a **masserator** rather than an accelerator because 99 percent of the energy it puts into a particle goes to increasing its mass rather than its speed. Electrons come out of the Stanford machine weighing more than protons, twenty times more!

When only one thing, in this case mass, can change as energy is added, the logic and analysis of the situation is easy. This is the ideal soft place to attack the problem.

Now the amount of energy put into something is the amount of force applied to it multiplied by the distance the force pushes it. Take, for example, the foot-pound, 1 pound pushing 1 foot. So

$$energy = force \times distance.$$

The amount of distance a thing moves is simply the time it spends traveling multiplied by its speed. So

$$energy = force \times time \times speed,$$

or

$$E = fts.$$

In the case of something traveling at nearly the speed of light, added energy can only add mass. If something traveling at a constant speed gains mass, a force is required to keep the increasing mass traveling at the constant speed. For example, if a conveyor belt is moving at constant speed while the mass of sand

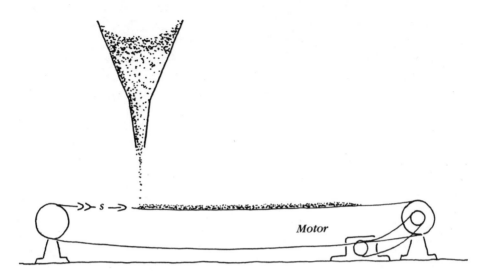

Figure 8–2.

riding it grows, then a motor is required to exert a force on the belt to keep it moving. Without the motor, even if the belt pulleys were frictionless, the belt would slow as the mass riding it increased.

If the rate at which sand falls on the belt is constant, the force necessary to keep the belt moving is proportional to the belt's speed. Double the speed and you double the required force. If the speed of the belt is maintained as a constant, the necessary force is proportional to how fast the mass riding the belt is increasing. Double the rate at which sand falls on the belt and you double the driving force required to keep the belt's speed constant. If the sand stops falling on the belt, the required force becomes zero, and the belt would coast by itself at constant speed were it not for pulley friction.

So the force required to keep the belt in motion is

$$force = speed \times rate,$$

or

$$f = sr.$$

The rate at which sand falls is expressed as

$$rate = mass / time,$$

or

$$r = m/t,$$

for example, grams of sand per second.

Take the force equation, erase the r and replace it with m/t, and you get

$$f = sm/t.$$

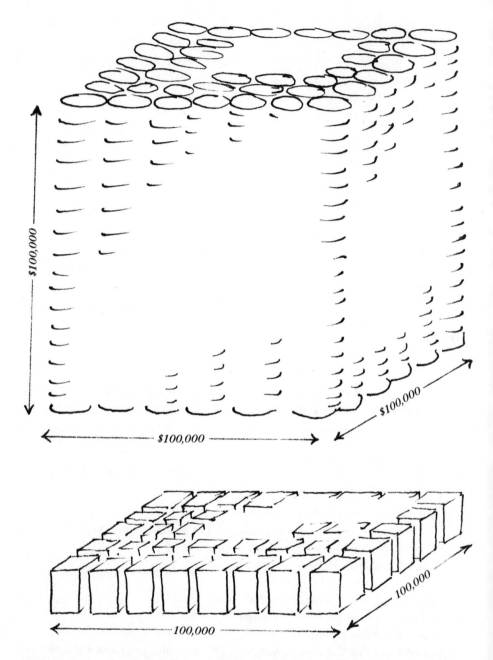

Figure 8–3. Visualizing big numbers. How do you picture big numbers? You can, for example, picture 1,000,000,000,000,000 = 10^{15} (it has 15 zeros) as a cubical stack of dollars. The stack is $100,000 wide (the price of a house in 1980), $100,000 high, and $100,000 deep. This is equivalent to the dollar value of a square city with 100,000 houses on each side.

Now take the energy equation

$$E = fts,$$

and erase the f and replace it with sm/t,

$$E = (sm/t)(ts)$$
$$\underbrace{}_{force} \underbrace{}_{distance}$$

Simplify this mess by canceling the two times and you get

$$E = mss.$$

The equation means that the *energy E* added to keep the belt in motion is equal to the added *mass m*, multiplied by the *speed s* at which the *mass* is moving, multiplied by the *speed* again, provided only that the belt's speed does not change during the story.

If a thing, like the electron, is moving at nearly the speed of light, its speed can hardly change, and that speed is called *c*. Only its mass can increase with added energy. So you have it in the bag,

$$E = mcc \text{ or } E = mc^2.$$

This equation tells you how much added energy there is in the added mass, or

$$m = E/c^2$$

tells you how much added mass there is in the added energy.

So what is the weight of a foot-pound of energy? With the value of *c* in feet per second and using appropriate conversion factors, it comes out that 1,000,000,000,000,000 foot-pounds of energy roughly weigh 1 ounce. Small wonder you notice no weight increase when you wind your pocket watch.

For teachers only. Why did God create light? God created light as an object lesson for physics teachers to demonstrate that their favorite shibboleth, *force* equals *mass* times *acceleration*,

$$f = ma = m \, dv/dt,$$

is not, as many unconsciously suppose, written on the back of the Ten Commandments. The equation found on the Ten Commandments is:

$$f = m \, dv/dt + v \, dm/dt.$$

This means that even if the *acceleration, dv/dt*, is zero, the *force* need not be zero if the *mass* of the moving thing changes with time. You can often get away with using this mistaken platitude. For example, it leads to this familiar kinetic energy equation, which goes as follows:

Since *energy* is *force* times *distance*, an energy increment is:

$$dE = f \, dx.$$

If $f = m\, dv/dt$ and $dx = v\, dt$, then

$$dE = (m\, dv/dt)(v\, dt) = m\, v\, dv,$$

integrating

$$E = mv^2/2.$$

This is the classical kinetic energy equation. However, it is good only when velocities are much less than the velocity of light, and mass is constant.

To demonstrate that the old saw $f = m\, dv/dt$ is less than the whole truth, God created light and made its speed constant. Since the speed of light can't change, dv/dt must always be zero. So hammering the $f = m\, dv/dt$ equation into students' heads as though it were the word of God is useless. For light the force equation becomes:

$$f = v\, dm/dt.$$

When you put this force into the energy increment equation, you get:

$$dE = (v\, dm/dt)dx = (v\, dm/dt)\, v\, dt.$$

The dt's cancel and the v we are talking about is c, so

$$dE = c^2\, dm,$$

integrating

$$E = mc^2.$$

This is the correct energy equation for light. It looks like the classical kinetic energy equation without the one-half factor.

To test your understanding of the situation in light of the above lesson (teachers love to test), ask yourself this: If a mass of sand m accumulates on the conveyor belt moving with constant speed v, how much energy goes into the sand, mv^2 or $1/2mv^2$? Hint: All the energy in the sand isn't kinetic.

A Popular Misconception

It was shown how the mass of an object can increase even though its speed does not, for example, winding a spring. It is surprising (even to some physicists) that the converse is also true. That is, the speed of an object can be increased even though its mass does not increase! There is a widespread misconception that mass always increases as speed increases.

Figure 8–4.

Take a motorcycle or better yet a wind-up toy, anything that carries its own source of motive energy with it. Set it in motion. As the toy runs it goes faster and faster. But no one adds any energy to the toy from outside; the toy had its energy in it from the start. If no outside energy is added, no mass is added. The toy merely converts the potential energy in its spring to the kinetic energy of motion. As you see it, the toy's total mass does not change.

But from the perspective of a passenger inside the toy the total mass does not even remain constant. It actually decreases! The passenger moving with the car regards the car as being at rest, and so cannot perceive the car's kinetic energy of motion. However, the passenger certainly does perceive the unwinding spring in the car, and as it unwinds it must lose mass. From the passenger's perspective the total mass, that is, the total rest mass or proper mass, must decrease even as the car's speed increases.

Contrast the wind-up toy or

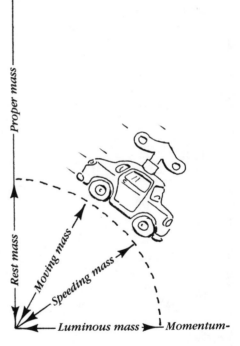

Figure 8–5.

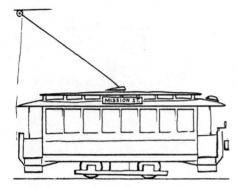

Figure 8-6.

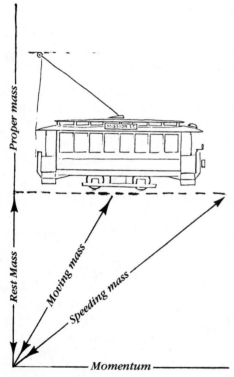

Figure 8-7.

motorcycle with a streetcar or an electron in the linear accelerator. Energy, and therefore mass, is poured into the accelerating streetcar or electron from an outside source, the local powerhouse. So you certainly see their mass increase in proportion to the energy poured in through the feed wires.

But again, the perspective of the streetcar passenger yields a different tale. The passenger moving with the streetcar regards the car as at rest, and so cannot perceive the car's kinetic energy of motion. From the passenger's perspective the mass of the car remains just the constant, unchanging proper rest mass.

Like everything else, this lends itself to a picture. As the streetcar accelerates its proper mass does not change, but its dynamic mass and momentum increase as illustrated. Even though its speed is limited by the speed of light, its mass and momentum are not. They can increase to infinity, just as they could in the good old physics before Einstein. But how can something moving at a finite speed, the speed of light, have infinite momentum? Because it has infinite mass.

As the wind-up toy accelerates its proper mass actually decreases. Its dynamic mass remains constant and its momentum increases, but not to infinity. The accelerating toy literally converts its proper mass into momentum, just as an accelerating clock converts its speed through time into speed through space. When all the toy's proper mass has become momentum it has transformed itself into light. Light has zero proper mass. Yet its dynamic mass, energy, and momentum are not zero. Were its proper mass larger than zero, its dynamic mass, energy, and momentum would all have to be infinite. Can you see why from the diagrams?

Is Mass Energy?

I have shown that energy always has mass. That is, adding energy to anything also adds mass. But does mass always have energy? For example, a thing at rest has mass—its proper rest mass. But it has no apparent energy. Might not there be two kinds of mass? One kind due to energy and the other just inert rest mass? Remember, the discovery that air had weight did not imply that all things with weight had air in them.

The Theory of Relativity contains no answer to this question. However, outside and quite apart from strict relativity, Einstein speculated that God would not bother to make two fundamentally different kinds of mass. Einstein speculated that energy not only made some mass (like the mass of light), but also that energy made all mass, that energy even made rest mass. But how?

I feel confident to venture that Einstein pictured pure energy—energy with zero rest mass—as light. And that he pictured the elementary particles of rest mass as two particles of light—photons—somehow

Figure 8-8. A particle of rest mass is pictured as two photons revolving about each other. This is a useful but incomplete model.

Figure 8–9. This sketch illustrates the disintegration of an atomic nucleus. The disintegration does not alter the elementary particles in the nucleus. It only alters the way they are connected to each other.

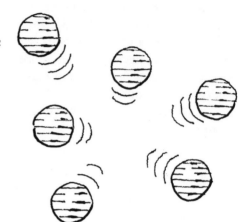

coupled and revolving. That way, their total momentum would be zero and their total angular momentum can be calculated to be the well-known angular momentum of elementary particles. If somehow the photons could be uncoupled, they would fly out in opposite directions and the rest mass would vanish into pure energy!

When in the 1930s energy was released from processes that reduced the mass of atomic nuclei (leading to atomic bombs and nuclear reactors), the notion was popularized that this verified Einstein's speculation on rest mass. In fact, it did not. The atomic nucleus contains various parts: protons and neutrons. These are held in place by the tension of opposing forces: the strong nuclear force of attraction and the electrostatic force of repulsion. This tension, like the tension in a spring, stores energy and therefore mass. The nuclear energy that was released came not from the rest mass of the protons or neutrons in the nucleus, but rather from the mass of tension energy in the "springs," which relaxed as the nucleus came apart.*

Nevertheless, Einstein's speculation on the fundamental nature of rest mass is probably very close to the truth. When the structure of elementary particles of rest mass is pictured, it will likely work out to be some kind of spinning vortex (as Lord Kelvin visualized) made not only of photons but containing all the things that move at the speed of light: gluons, gravitons, neutrinos, etc., so mixed as to provide their own self-coupling.

---

*The later discovery that for every particle there exists an antiparticle that can reduce the particle and itself to pure energy may be taken as much better experimental verification that rest mass is made of energy.

E = mc² by bouncing light from a moving mirror. This box shows how to establish the relationship between the energy and mass of light, $E = mc^2$. In familiar language it shows how to calculate the number of pounds in a foot-pound of light energy, $m = E/c^2$.

Most light gets its energy (and mass) from atomic electrons, but how the process works is not easily explained,* so let's avoid it. How can you directly put a foot-pound of energy into light? By pushing energy into it. Light exerts force on things it hits (remember the comet's tail), which means that light has mass and momentum. Suppose the light exerts a force of 1 pound on a mirror. If the intensity of sunlight could be increased 10 million times, it would exert a force of roughly 1 pound on a mirror 1 foot square. Suppose you push the mirror 1 foot against the 1-pound radiation pressure force. Then you have pushed 1 foot-pound of energy into the reflected light. That is, the light bounces back from the mirror with all the energy it had when it hit, plus one extra foot-pound of energy. Now I will show how to reason out what the mass of that added energy is.

How does a mirror work? I don't know. But when I was a little boy my mother told me there were things behind the mirror that came out. (One of the things was a black hand that would come out and choke you if you made faces into the mirror.) After a while I decided (since the black hand never came out) that the things behind the mirror only tried to come out, but were pushed back by things in front trying to go in. That explained reflection!**

That idea is illustrated in the first sketch. Some light of mass M and momentum Mc tries to enter a stationary mirror. It is repulsed by a similar mass of light with opposite momentum approaching from behind the mirror. As the masses exchange momentum there is a moment when both come to rest and then they fly off in opposite directions. Note that it does not really matter whether you take my version of the

---

*It is hard to understand in large part because of the failure of quantum electrodynamics theoreticians to come up with a good, easily understandable working myth.

**In the quantum electrodynamic version of the story, the mirror works because the atomic electrons in the mirror's silver capture the mass of light that hits them, momentarily hold the mass prisoner, and then expel it in the opposite direction. In all versions of the story, what is behind the mirror (the wall, someone pushing, or imaginary light) is required to provide the punch, the momentum, necessary to throw back the incoming light. Moreover, it is vital to appreciate that exactly what is behind the mirror is irrelevant. The only thing that is relevant is that it provides the required momentum.

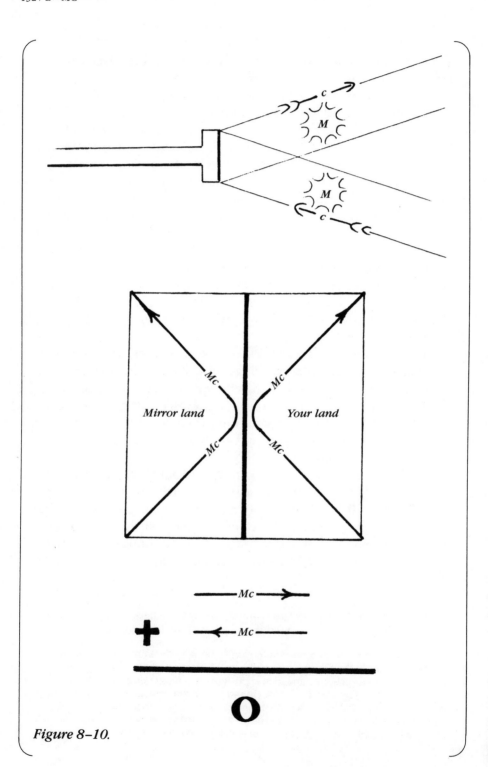

Figure 8–10.

story (momentum exchange) or my mother's (thing behind comes out), since in each case the momentum that was behind the mirror ends up in front and the momentum in front ends up behind.

Now suppose the reflecting mirror is not stationary but advancing toward the oncoming light with velocity v. The advancing mirror must increase both the momentum and energy of whatever it reflects, just like a tennis racquet that has whacked a ball. (A receding mirror will decrease both the momentum and energy of what it reflects and a stationary mirror will simply reverse the momentum and leave the energy unchanged.)

The momentum and energy of the whacked tennis ball are increased by increasing its speed, but the speed of light can't be changed. So how can the momentum and energy of the reflected light be increased? By increasing its mass. (That is the only thing about the light that can be changed.) The mass of light bouncing from the advancing mirror is the mass M that hit it, plus some extra mass m that came from whatever pushed on the reflected light (in my story the light behind the mirror).

The sketch shows the mass of light M approaching the front and $M+m$ approaching the rear of the mirror. Now, according to Mom, $M+m$ pops out and M drops into the mirror.

According to me, $M+m$ tries to get out but is hit by M. The M trying to enter the mirror reflects from the M trying to come out of the mirror and sends it back into the mirror, but nothing stops the m and it escapes out of mirrorland.

Or you can suppose that $M+m$ hits the receding back side of the mirror and is reflected back into mirrorland with only mass M, while M hits the advancing front side of the mirror and is reflected forward with the enhanced mass $M+m$.

In all versions of the story the light behind the mirror loses mass m and the light in front gains m. That is, the light reflected from the advancing front of the mirror gains the energy of mass m, and whatever agency pushes the mirror from behind puts out the energy of mass m.

Both before and after reflection, the momentum of the light going to the right is $Mc+mc$ and that going to the left is $-Mc$. So the net total momentum is $Mc+mc-Mc=mc$. At the instant of reflection (momentum exchange) all mass must momentarily come to rest in the mirror, that is, it must be moving with the mirror. So at that instant the net total momentum must be the sum of the mass of the incoming and outgoing light multiplied by the mirror's velocity, $(M+M+m)v$. So: $(M+M+m)v = mc$. Store this fact for recall.

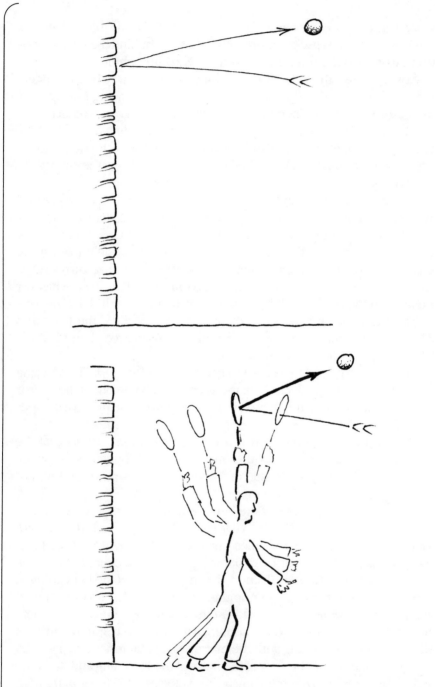

Figure 8–11.

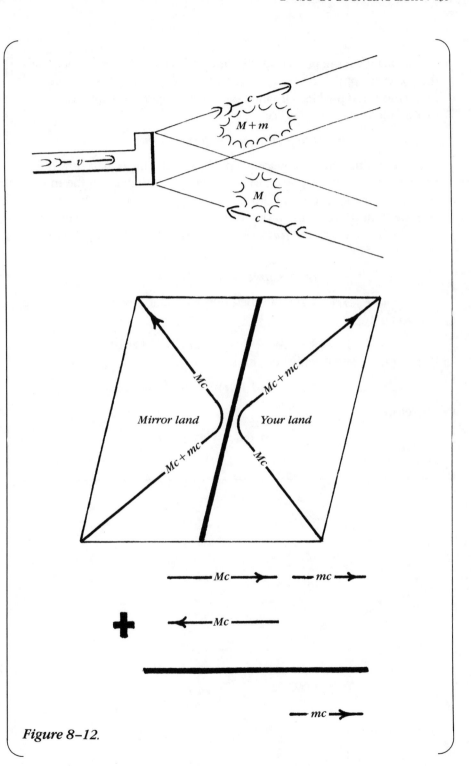

Figure 8-12.

The *energy E* the reflecting mirror adds to the reflected light equals the *force* it applies (radiation pressure force) multiplied by the *distance* the force pushes. Distance equals *time* spent moving multiplied by the mirror's *velocity v*. So:

$$E = force \times distance = force \times time \times v.$$

Force multiplied by how long it pushes on a thing equals how much momentum or punch goes into the thing. The momentum the mirror (or imaginary light behind the mirror) soaks up in stopping the on-coming light, *Mc*, plus the momentum it pushes into the reflected light, *Mc + mc*, must equal the *force* on the mirror multiplied by the *time* it acts. So:

$$force \times time = Mc + Mc + mc.$$

Now replace *force* × *time* in the energy equation with *Mc + Mc + mc* and you get

$$E = (Mc + Mc + mc)v = c(M + M + m)v.$$

Recall it has been established that:

$$(M + M + m)v = mc.$$

Therefore,

$$E = cmc = mc^2.$$

Amen.

> The ways of nature are mysterious, but they are rational, taken all together.

Chapter 9
Enigma

Light Falls

If light carries energy (and it does, for how else could it give you a sunburn?), it has weight.* And if it has weight it should fall.

The notion that gravity might cause light to fall can be traced back to the astronomer Simon Laplace. The year Napoleon was commencing his foreign wars the astronomer was analyzing the effect a star's gravity might have on the light escaping

Figure 9–1. At the age of twenty-five, Soldner published an article "On the Deviation of a Light Ray from Its Rectilinear Motion, Through the Attraction of a Celestial Body Which It Passes Close By." Soldner was born the year the North American colonies commenced their insurrection against Great Britain. He lived in Bavaria, the same area Einstein came from, which is now southern Germany.

---

*In the preceding chapters the words mass and weight were used interchangeably. The implication was that if energy has inertial (or momentum) mass it also has weight, which is hardly surprising since there is only one kind of mass. If there were two kinds, perhaps one kind could have inertia without weight. However, in spite of several deliberate searches, no one has ever found mass that did not have both inertia and weight in equal proportions.

from it. In 1800 the geodesist (map-maker) Johan Soldner calculated by how much light passing the sun or the earth would be deflected by gravitational attraction. It might even be suspected that light could fall if it had no mass at all, since (as Galileo showed at the Leaning Tower of Pisa) the motion of a falling body does not depend on its mass. Yet, this line of physical thought became dormant until its rude 1906 reawakening by the realization that all energy (including light) carries mass—the same kind of mass—that mass has weight, and that weight falls. However, this very realization, which his own mass-energy relationship made unavoidable, led Einstein into a deep dilemma.

Can C Change?

The sketch shows the (exaggerated) deflection of a flashlight beam by gravity. The wave front that starts at AC on the flashlight ends up on the screen at BD. While the light on the top of the beam travels from A to B the light on the bottom travels from C to D. The light on the bottom travels the inside of the curve (the distance CD is shorter than AB) and so has a shorter path from flashlight to screen. The wave front that started from position AC eventually comes into position BD. That means that it took as much time to go from A to B as to go from C to D, even though the path from A to B is longer than the path from C to D. That means that the light had to

---

Heavy light. In 1800 many respected scientists thought light could be deflected by gravity. But then the thought was not pursued for well over a century. Why was it abandoned? Because in 1800 many respected scientists also thought light was made of little pellets. In the following decades that idea of light came to woe and the wave theory of light gained ascendancy.

Waves (like sound or earthquake waves) have no mass; they don't even have momentum. So there was little reason to believe gravity should affect them. Halfway through the century Maxwell showed that light waves, unlike most other waves, do carry momentum. But few contemporaries took him seriously, and no one insisted momentum must imply mass until Einstein did in the first decade of the 1900s. That reawakened the old idea that light had weight and should therefore fall.

Incidentally, Einstein pushed back yet further along this road of thought and concluded that light not only carried mass but also carried it in little pellets! The pellets were called corpuscles by Newton and rechristened photons by Einstein. The photons became one of the cornerstones of Quantum Theory.

travel slower* along the lower path! But the central idea of all of Einstein's work was that the speed of light is constant. Thus Einstein fell into the grip of a powerful enigma, an enigma of his own making. Perhaps his theory was not self-consistent—perhaps it was irrational.

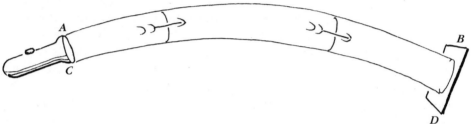

Figure 9–2.

---

*It was shown in a question at the end of the first chapter that a light wave must always move in a direction perpendicular to itself and cannot move obliquely. If the speed of light traveling from A to B was the same as the speed of light traveling from C to D, then the light would arrive at D before B and the wave would be forced to travel obliquely—which it cannot do.

If a light wave did move obliquely, how would you know it? You would know it because it would have a longitudinal component so you could not polarize it completely.

Figure 9–3.

Crisis

To fall, a light beam has to turn down. To turn down, its lower side has to move more slowly than its upper side. So the speed of light near the earth or in any gravity field has to decrease and keep on decreasing as the light goes lower in the field.

Not only did this trouble Einstein, but also his detractors had a field day. Some people disliked him. (Even his wife disliked him.) Who likes a young upstart with the audacity to jack around with space and time? Some people disliked his theory. Who likes a theory that is difficult to grasp no matter how much physics you might know? Certainly a professor who had mastered everything except the new theory would not love it. Now there seemed to be a flaw in the keystone of Einstein's thought bridge, and there were people outside and inside physics who would be relieved to see it crash.

But Einstein was a most cunning thinker. He came up not only with a clever way to get out of the enigma, but also in so doing came up with an explanation of the cause of gravity!

Slow Time Again

Einstein cured the paradox by a means typical of his previous inventions. He said that it was not the speed of light but the speed of time itself that runs ever slower as you descend into a gravitational field.

This theory automatically answers the speed-of-light problem because if time runs slowly, then everything, including the speed of light, slows down. But what does it gain? That is, why not just simply admit that light slows down and leave it at that? The difference between just slowing light and slowing time is that if time is slowed, local observers down in the gravitational field, whose time is slowed, will measure the normal, unchanged speed of light; that is, they will be faked out. Distant observers who are remote from the gravitation will not have their time slowed and so will measure the light in the region of gravity being slowed down.

But jacking around with something as fundamental as the speed of time will not only affect the speed of light. It will also have all kinds of other consequences. For example, suppose the twins Peter and Danny are exactly the same age. Peter climbs up to the attic and Danny goes down to the basement. After a while Peter drops back down and Danny comes up. The twins are back together but Peter is older than Danny! Time in the basement was running slower than time in the attic.

In fact, the twins don't even have to get back together. They can just send messages to each other. Any kind of message by any kind of messenger. For example, heartbeats sent by a tin-can telephone. Peter would hear Danny's heart beating more slowly than his and Danny would hear Peter's beating more quickly than his.

The frequency change has been experimentally detected by comparing atomic clocks at the top and bottom of towers. It also reduces the frequency of light that comes up through a strong gravity field and so produces a detectable gravitational red shift in the light escaping from very massive stars.

The time slowing produced by gravity differs from the time slowing produced by motion, in this way: If the time slowing is produced by motion, I think my clock is OK but yours is slow, and you think yours is OK but mine is slow. If the time slowing is produced by gravity and you are nearer the earth's surface than I am, then I think my clock is OK but yours is slow, and you think yours is OK but mine is fast.

Curved Spacetime

I will now show how the myth and the diagram by which the myth is visualized can be generalized to produce slow time and, most astonishingly, how the slow time in turn produces the effect you call gravity.

Suppose you take a spacetime diagram and roll it up as illustrated. Rolling it in no way affects the diagram itself. Drawings are always rolled up when put into mailing tubes and inside the tube lives the same drawing that lived outside. You might have a harder time reading

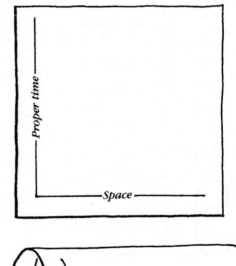

Figure 9-4.

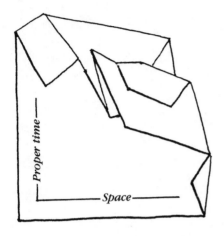

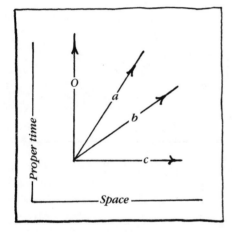

Figure 9-5.

Figure 9-6.

the rolled drawing, but some machines (facsimile scanners) can read it more easily when it is rolled. In a whimsical way, it is appropriate to roll the diagram so that time can run around a circle, as it does on a clock.

If you wanted to be brutal, you could even crumple the diagram without affecting the drawing it carried. The pattern of a dress is not altered by crumpling it. That is, the same thread still joins the same points. The fashion of the dress is internal to its surface and does not depend on its shape in three-dimensional space. Only cutting the diagram or dress will alter the intrinsic way its parts are interconnected.

When the diagram is rolled up, objects that move straight through spacetime will spiral down the cylinder, slower-moving objects forming tighter coils. Stationary objects, traveling only through time, will simply circle the cylinder. Light beams, moving only through space, will go straight down the cylinder. This diagram, rolled or unrolled, represents spacetime without gravity.

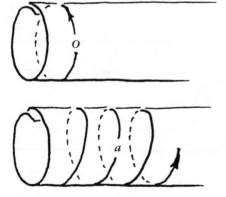

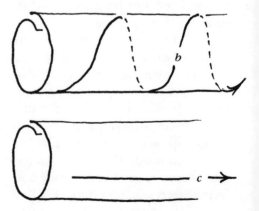

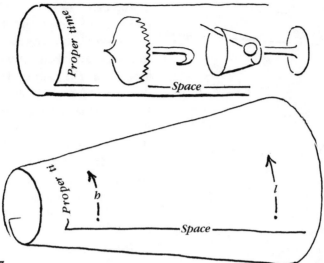

Figure 9–7.

Now I will alter the diagram to show the effect of gravity on time. Suppose the space direction runs vertically down into a gravity field, as shown by the orientation of the umbrella and wine glass. What can be done to make objects that are at lower locations age more slowly than objects above them? The cylindrical spacetime diagram can be flared into a cone, wide end down.

Let there be two stationary objects, one above the other. Each one travels only through time. Accord-

ing to the myth, all objects move through spacetime with the same speed. Let each object be represented by an ant. The ants crawl at the same speed. They crawl around the cone in the time direction. But the high road *b* around the narrow upper neck of the cone is shorter than the low road *l* around the base of the cone. So the ant on the high road gets ahead of the ant on the low road. Ahead means ahead in proper time or age.

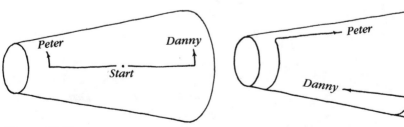

Figure 9–8.

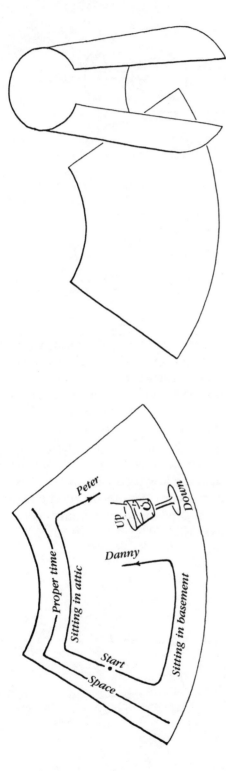

You can now graphically see what happens to Peter and Danny. The twins start from the same point at the same age. Peter goes up to the attic and sits there. Danny goes down to the cellar and sits there. Each travels through spacetime at the same speed; that is, each moves the same distance on the diagram — moves through the same amount of coordinate time. But when they decide to get back together Peter finds himself ahead of Danny. Ahead in age.

To see the whole story in one view, slit the cone open and lay it flat. It is apparent that the length of the path traveled by Peter is equal to the length of the path traveled by Danny. The path length measures the amount of lapsed coordinate time. But Peter ends up ahead of Danny in proper time.

Coordinate time is the measure of duration perceived by you who are observing from a position in distant space outside the influence of gravity. Proper time is the measure of duration perceived by the thing itself under the influence of gravity. Proper time is personal time or age. Thus, the conical spacetime diagram does what I promised it would do: produce slow time where it is needed.

Figure 9–9. It looks like Peter's path might be longer than Danny's, but that is an illusion. Measure both path lengths and see for yourself.

How slow? Gravity slows the speed of time. In this box it will be calculated how much time must be slowed down to give a light beam the downward acceleration g. (The downward acceleration of light is the same as the downward acceleration of all mass). The top and bottom of a flashlight beam are separated by a vertical height h. When some amount of time T goes by at the top of the beam, a lesser amount $(T-t)$ goes by at the bottom. I will calculate the pure ratio t/T. (Note: 100 t/T is simply the percent discount in the speed of time. The discount might go to 100 percent in very, very extreme and unusual situations, in which case time stops running.)

While the top of the beam moves a distance cT, the bottom moves the lesser distance $c(T-t)$; that is, the top of the beam moves a distance ct farther than the bottom. This tips the beam and its direction of motion downward. After a time T, the downward component of the beam's velocity must be the same as the downward speed of any falling mass gT. Now the two little right triangles in the sketch are similar, so the ratio of ct to gT is the same as the ratio of h to c. Thus,

$$ct/gT = h/c,$$

so

$$t/T = gh/c^2.$$

Note that the little triangles are similar or proportional because each has the same angles. Each has the same angles because ct is parallel to c and gT is parallel to h.

If h is your height, about 6 feet, and g is the acceleration of the earth's gravity, 32 feet per second per second, the speed of time at your feet works out to be about 0.00000000000001 percent less than the speed of time at your head. That tiny difference, as you shall see, is not only sufficient to bend light, it is sufficient to cause gravity itself!

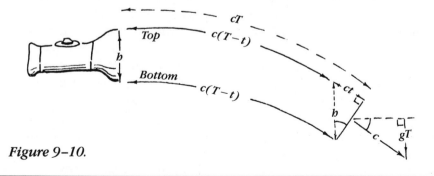

Figure 9-10.

How does the earth exert force
on the moon when there is no
connection, but only empty
space, between them?

Chapter 10

The Cause of Gravity

The Origin of Gravity

How does the force of gravity get
to an apple? After all, no visible
string pulls it earthward. The agency
that acts on the apple must be invisible. What is the most familiar invisible agency you know of that can
exert force? The wind. So Galileo
first thought a wind blowing from
space down onto the earth was the
cause of gravity. Quite literally, the
earth sucked. This is why Galileo
first (erroneously) thought all falling
objects reached a terminal falling
velocity (the speed of the wind) and
then held that constant speed the
rest of the way down.

However, if you ask yourself a
series of questions about being hit
on the head by a heavy object falling
1 inch or 2 inches, 1 foot or 2 feet,
1 yard or 2 yards, you will (as Galileo did) quickly come to realize that
the speed of a falling object continuously increases as it falls. Thus, Galileo next concluded that the speed
of a falling object is proportional to
how far it falls. So the wind idea had
to be discarded and there was no
longer a nice picture of the cause of
gravity.

Furthermore, Galileo was still
wrong. The speed of a falling object
is not proportional to how far it falls.
It cannot be. To see why, suppose an
object has a certain speed after falling 1 foot. Then it would have half
that speed after falling $\frac{1}{2}$ foot, and a
tenth of that speed after falling $\frac{1}{10}$
of a foot. To get any speed it would
have to go down a little. But if it
started with zero speed, how long
would it take to go down even a
little bit? It would take forever! So
an apple would not begin to fall by
itself. If you gave the apple a start by
pushing it down, it would get going
and begin to fall faster and faster, but
the apple would not be self-starting.
Now the world might have been
made so that apples would not fall
by themselves. After all, gasoline engines and old electric (clock) mo-
tors are not self-starting, but apples
are self-starting, at least in this
world.

Finally, Galileo decided that objects gain their speed not by virtue of passing through space, but by virtue of passing through time. So, even if the apple started with zero speed, and did not (to begin with) pass through space, it could still get speed because it passed through time, and the gained speed came from passing through time. The speed of a falling object is proportional to how long, not how far, it falls.

But how could passing through time cause an object to gain speed? It was a rule without a reason. There had to be a reason (in physics there must always be a reason).* Reason, like speed, is gained by passing through time. In this case it took three centuries of time. And when it came it was a shocking surprise, because it was found that the falling speed did not come from strings, or wind, or pressure, or any other kind of force—it came from the very fashion (shape) of spacetime itself.

---

*Physics is actually a religion. A religion is based on articles of unproved faith. In physics the articles of unproved faith are: "There always has to be a reason" and "The body of reasons has no self (internal) contradictions." If past experience had not contradicted these articles, that in itself could not assure continued compliance. But past experience has not confirmed the two articles. At any time, past or present, there are always things for which reasons have not been invented and things haunted by internal contradictions. Thus, belief in the articles is unproved faith.

Surprise!

The curved spacetime diagram was expected to produce slow time, and it did. But now I will show you something you won't expect, something remarkable. Go back to the zero-gravity cylinder diagram and draw the path of a stationary object. You can make the path begin at some point A and run on through time. The path just wraps around the cylinder. It does not move to the right or to the left; it is stationary in space.

Now flare out one end of the cylinder, making it into a cone. On the cone draw the path of an object that starts at rest and is then left to itself. The path of an object left to itself is a straight line. How do you draw a straight line on a cone? The same way you do on a cylinder, lay the cone out flat and draw a straight line on it. At point A the object is stationary, so its path is not tipped right or left, but runs straight through time. Just extend this line with a ruler and you have the path. Now look where the line goes. Little by little, it inches toward the wide end of the cone. The wide end is the slow time end, the down end. Lo and behold, the stationary object does not remain stationary. At first quite slowly, but then ever faster and faster, the pitch of its path increases. Down it falls. The conical surface causes stationary objects to fall. Slow time causes the spacetime surface to be conical. You have just uncovered the cause of gravity.

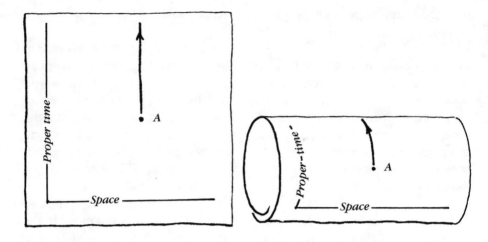

Figure 10–1.

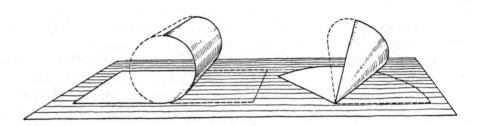

Figure 10–2. Development of cylinder and cone.

Slow time was introduced as a weird side effect of gravity. But now you can see that slow time is the very cause of gravity!

At this point, I strongly suggest that you close this book, take paper, pencil, ruler, scissors, and tape in hand and make some paper cylinders and cones and see and feel how they work. Do it now.

How Fast Can You Fall?

Not only is the cause of gravity revealed but also you have a bonus. A problem with the conventional idea of gravity is that the speed of a falling object always continues to increase. So what is to stop a falling body from accelerating past the speed of light? Its increasing mass?

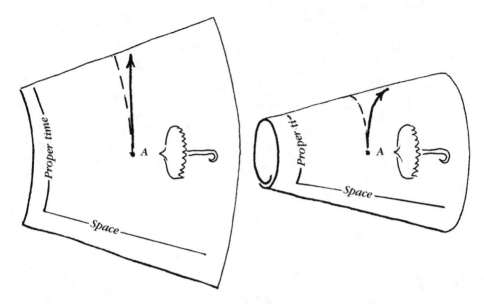

Figure 10–3. *The angle between the tangent line and the circular arc starts from zero at* A *and then grows from there in proportion to how far you move along the arc. Since time runs in the direction of the arc, and since you move continuously through time, it follows that the angle between the tangent and arc grows in proportion to time. Thus the downward pitch of the tangent line on the cone is proportional to time, and the pitch of the line is proportional to the downward speed of the falling body, which started from rest at* A *So the downward speed of the falling body grows in proportion to time, which is Galileo's rule for falling bodies. This is good so long as the falling speed is much less than the speed of light, which means so long as the angle between tangent and arc is small.*

No, that can't be the answer because the acceleration of a falling object is independent of mass.

The answer is in the diagram. The speed of the falling object, like the speed of anything else, is constant through spacetime. If the falling object starts from rest, it starts moving only in the direction of time. And it continues moving in the same direc-

tion at the same speed. Its travel leads it into another part of the spacetime diagram. In the other part, the direction of space and the direction of time are different. The motion originally directed through time finds itself directed through space. In the extreme case, when the path is infinitely long (that means after an infinite amount of

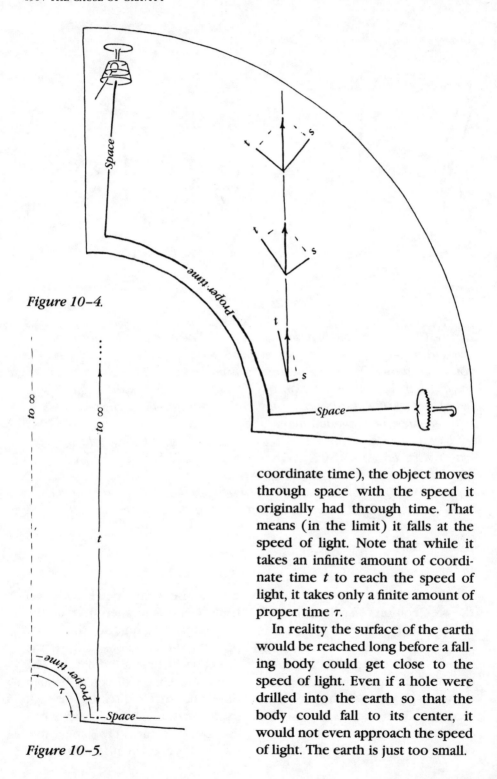

Figure 10–4.

Figure 10–5.

coordinate time), the object moves through space with the speed it originally had through time. That means (in the limit) it falls at the speed of light. Note that while it takes an infinite amount of coordinate time t to reach the speed of light, it takes only a finite amount of proper time τ.

In reality the surface of the earth would be reached long before a falling body could get close to the speed of light. Even if a hole were drilled into the earth so that the body could fall to its center, it would not even approach the speed of light. The earth is just too small.

The floor comes up. The conical spacetime diagram causes stationary objects to fall downward, but how do you know they fall with the acceleration of gravity, one *g*? To see that they fall correctly, turn the cone so that down is down, slice it open, and unroll it. Now consider the ceiling and floor of a house. The arc lengths of the cone at these locations must be in inverse ratio to the speed of time (and the speed of light) at these locations. That is, the longer the arc, the slower the time. Thus, the arc's curvature must be the inverse of the curvature of the falling flashlight beam. That means the arc's curve up is the same as the beam's curve down. The beam's downward acceleration was one *g*, so the arc's upward acceleration must be one *g*. Thus, an object traveling along a horizontal straight path has an effective downward acceleration of one *g* relative to the arc.

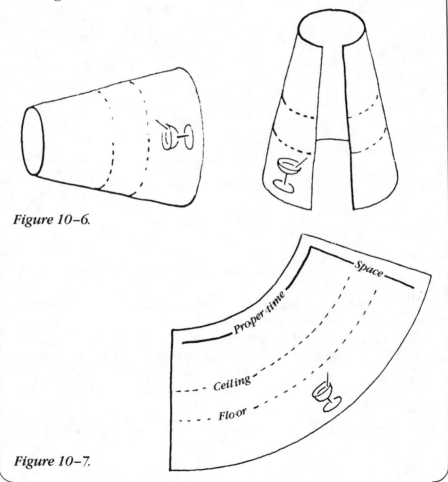

Figure 10–6.

Figure 10–7.

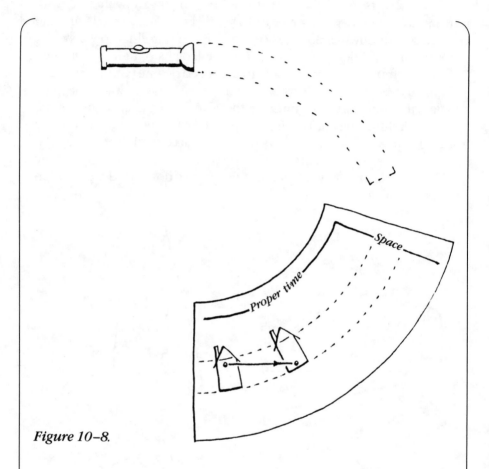

Figure 10–8.

If you wanted to interpret the unrolled spacetime diagram literally, you could say the house has an upward acceleration of one *g*. In a nutshell, Einstein's view of gravity is that things don't fall; the floor comes up! That easily explains why heavy objects don't fall faster than light objects. But don't take it too literally, because if the floor is coming up in both New Orleans and Calcutta, the earth's diameter could not remain 8000 miles.

Galileo's explanation. The following is Galileo's explanation of why large and small masses (disregarding air resistance) fall at the same rate.

The acceleration of a large falling rock is the same as the acceleration of a small falling rock because a large rock is just a bunch of small rocks falling together.

This explanation is occasionally reinvented by people who think about these things. Though not the first to think this way, they walk in the footprints of the old master!

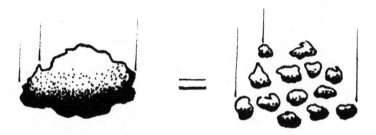

Figure 10–9.

A Hole to China

Suppose the hole was drilled into the earth. Suppose it was drilled clear through the earth's center and out the other side—the proverbial hole to China. (If the hole were drilled from North America it would emerge in the Indian Ocean near Australia.) I want you to think about the hole, because it illustrates the mechanics of the spacetime diagram in passing from a region of almost zero gravity in distant space to a region of strong gravity at the earth's surface and back into a zero-gravity region at the earth's center. Remember, inside the earth gravity decreases, becoming zero at its center.* If you doubt that gravity is zero at the center, tell me, which way would you fall?

The fashion of the spacetime diagram for a spaceline running through a hole in the earth can be reasoned out as follows: As usual, proper time is measured in a clock-like manner (angularly) around the diagram. Coordinate time is simply the linear length of the track left by the moving body on the diagram. And space is measured left and right **along the surface** in the direction

---

*Don't confuse gravity with pressure. As you go into the earth, pressure increases even though gravity decreases.

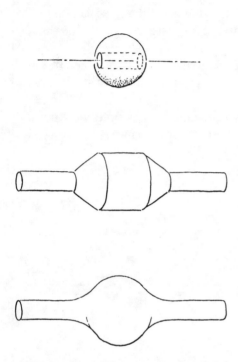

Figure 10–10. If an object falls straight down a well, then as far as that object is concerned, space has only one dimension and it runs straight through the earth. The spacetime surface for such a one-dimensional space is illustrated above. The surface made of cylinders and cones is a crude first approximation of the more accurate smooth surface below it.

The point of inflection between sphere and cylinder, where the spacetime surface is steepest, is where gravity is strongest. That place is at the earth's surface. The transition from sphere to cylinder should extend to infinity, but most of it occurs near the sphere, as sketched.

of the line through the hole. In the zero-gravity region far from the earth, the spacetime diagram is just the cylindrical surface previously discussed. Near the earth's surface where strong gravity comes into play, the cylinder becomes a cone, also previously discussed. At the earth's center, where gravity is again zero, there is another cylinder. Going out the other end of the hole the story is the same. The transition from cylinder to cone, like the transition of the gravitational force it represents, is gradual rather than abrupt. So the sharp edges of the spacetime surface are smoothed out. The smoothed-out spacetime surface inside the earth (inside the hole) is spherical. But for heaven's sake, don't confuse this spherical surface with the sphere of the earth!

What is the spacetime surface good for? It shows how objects can move along the line through the hole. First take an object resting at the earth's center. Its path on the spacetime surface is just a track around the diagram's "equator." That is, it does not move left or right through space, it moves only through time, as shown in Figure 10–11.

Trapped Inside

Next, suppose you give the object a gentle nudge to the right. It would begin to move slowly to the right. This you would represent by tilting the path of the object on the spacetime diagram slightly to the right. Now the object would track around the spacetime surface on a great circle inclined to the "equator." The track would resemble the track of the ecliptic on a globe (Figure 10–12). If you laid the spacetime surface flat (which is hard to do—have you ever tried to lay an orange peel flat?), the "equator" would become a straight line through time and the "ecliptic" would wobble right and left (like a sine wave) around the straight timeline. What is this telling you? It is telling you that the object you nudged moved a way to the right until gravity stopped it. Then gravity pulled it back. But there was nothing at the center to stop it, so it overshot the center. Eventually, gravity stopped it and hauled it back toward the center again. It falls back and forth in simple harmonic motion, like a pendulum.

Newton's view of this affair would have been that the force of gravity exerted a springlike force on the object, causing it to oscillate back and forth. Einstein's view would have been that there was no spring or any other force. The object just moved straight (or as straight as it could) along the spacetime surface. If that surface was curved, it would curve the track of anything moving on it.

Next, suppose you gave the object a harder nudge to the right. It would begin to move rapidly to the right. This you would represent by tilting the path of the object on the spacetime diagram strongly to the

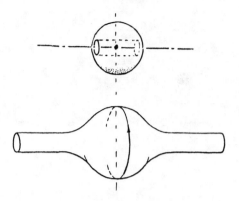

Figure 10–11.

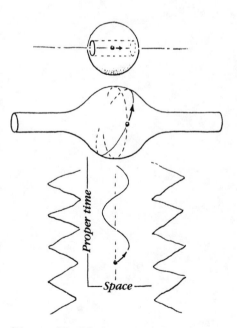

Figure 10–12.

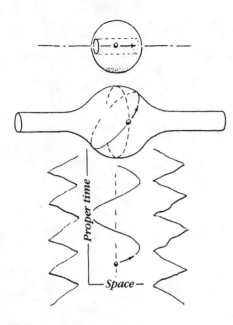

Figure 10–13.

right (Figure 10–13). Now the object tracks around the spacetime surface as it previously did on a great circle, but the inclination of that circle to the "equator" would be increased. Thus, the object would oscillate back and forth with a larger amplitude, that is, it would get farther from the center before it returned.

Now I will show you a little curiosity. If two objects start together at the center of the earth, and if one is kicked harder than the other and so goes higher before falling back, they will nevertheless both arrive back at the center together. The reason for this phenomenon is revealed by the track of the oscillating object on the spherical spacetime surface (Figure 10–14). The circumference of a sphere, whether measured along the equator or the ecliptic, is the same. It does not even matter how much the ecliptic is inclined. The length of the track around the spacetime surface measures the period of oscillation. So the object in the hole is like a pendulum (Figure 10–15). When it swings farther it also swings faster; thus, the time for a round trip does not change.

Question

Suppose the identical twins, Peter and Danny, are the same age and at the center of the earth. Peter stays at the center, but Danny pushes himself off and begins to fall back and forth in the hole. That means Peter remains put while Danny goes on excursions, like a pendulum, away from and back to Peter. So Peter is the stay-at-home twin and Danny is the traveling twin.

a After a while, it is found that Peter has aged more than Danny.

b In this situation, the stay-at-home twin and the traveling twin find they are still the same age at reunion.

Answer

The answer is **b**. If the spacetime surface was flat, the traveling twin would come to reunion less aged than the stay-at-home twin. But in the earth's field of gravity the spacetime surface is not flat. Every half cycle, that is, every time they go 180 degrees around the spacetime surface, Danny swings back to Peter. These reunion points are on opposite sides of the spacetime sphere, so thay are antipodes. Peter and Danny both travel great circle paths between the points, so their paths between antipodes are of equal length, and therefore they arrive at the antipodes simultaneously. With the help of gravity, the twins return to the same place, at the same time, and at the same biological age. Remember, degrees of turn around the surface measure age, and both twins go through 180 degrees between reunions.

If the twins were out in empty space, and Peter stayed put while Danny bounced back and forth on

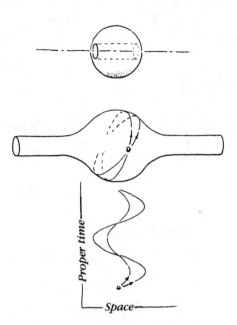

Figure 10–14. All great circles on a sphere have the same circumference. Therefore, all objects require the same time to oscillate in the hole.

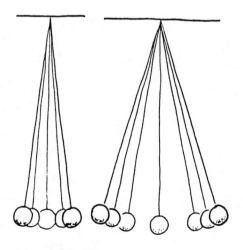

Figure 10–15.

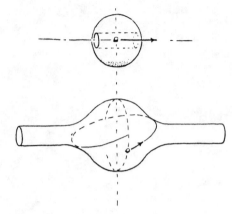

Figure 10–16.

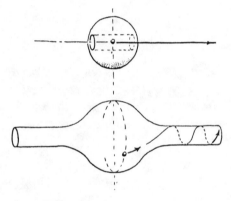

Figure 10–17.

a long spring, and gravity was not involved, then at reunion Danny would be biologically younger than Peter. You can now begin to appreciate that gravity is more than a force.

Escape

However, if the object goes high enough to emerge from the hole, the story begins to change. The object's track passes off the spherical part of the spacetime surface (Figure 10–16). Thus, the length of the round-trip track increases, which means that the object takes more time to do its round trip.

If the object is kicked hard enough, it will not only get out of the hole but will also go high enough to get onto the cylindrical part of the spacetime surface. Once on the cylindrical part, there is no falling back. The object spirals on forever. The object has escaped (Figure 10–17). Remember that the unrolled spiral track is a straight line through spacetime. The key thing is the path of the track on the surface, not the (spiral) path of the track through three-dimensional space.

Fashion, Not Shape

It is important to understand that the path of the track is controlled by the fashion of the surface itself, that is, by the way the threads that make the surface are interconnected among themselves, and not by the three-dimensional shape of the surface. The shape is chosen to make a good display of the surface, to make the teaching easy, just as a dress is displayed on a manikin to make the selling easy. In a washing machine the three-dimensional shape of the dress is twisted and mangled, but the fashion of the dress, though hard to recognize, is not changed in the least. Just as a dress can be laid flat by either folding it or crumpling it,

so also can the spacetime surface be flattened and still remain the same surface with the same tracks. The sketch in Figure 10–18 shows a simplified spacetime surface, slit and unrolled. The part inside the earth (the wide part) has fluted corrugations when unrolled. These can be crumpled over flat, if you insist. Except for the edges, blind ants crawling on the surface could not distinguish between the rolled and unrolled surface.

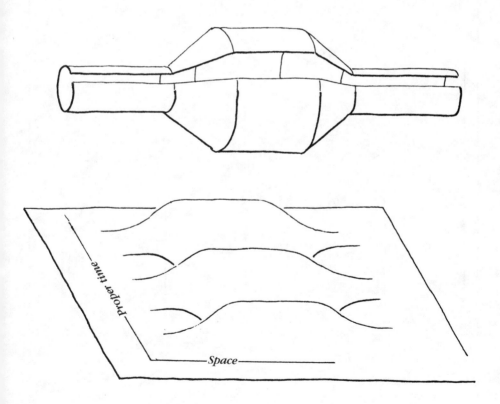

Figure 10–18.

Gravitation by refraction. If you object to curved spacetime, just about all gravitational effects can be pictured without curved spacetime if you will let the speed of objects through spacetime be variable rather than constant. The speed decreases in and near the earth. Visualize a fiber-optic light pipe running through time. The earth moves through time at the center of the pipe (Figure 10–19). The refractive index of the pipe decreases away from its center, as is indeed the case with real light pipes. The speed of objects through spacetime is inversely proportional to the index of refraction of the region they are traveling through. The greater the index, the slower the speed.

Refraction in regions of variable index is best illustrated by a mirage. Remember, the ray's path is bent toward regions of high index and away from regions of low index (Figure 10–21).

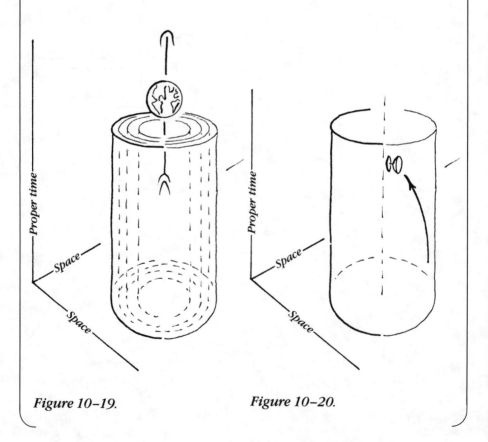

Figure 10–19.   *Figure 10–20.*

Figure 10–21.

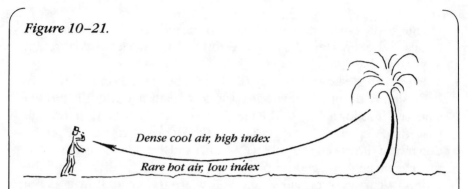

Dense cool air, high index

Rare hot air, low index

Applied to motion through spacetime, the path of a goblet that was initially stationary in space, moving only through time, is "refracted"— accelerated toward the earth.

Figure 10–22 shows a stone in a hole through the center of the earth being "refracted" back and forth. This is how light goes down a light pipe. Figure 10–23 shows a spacecraft orbiting the earth. This is another mode for light going down a light pipe.

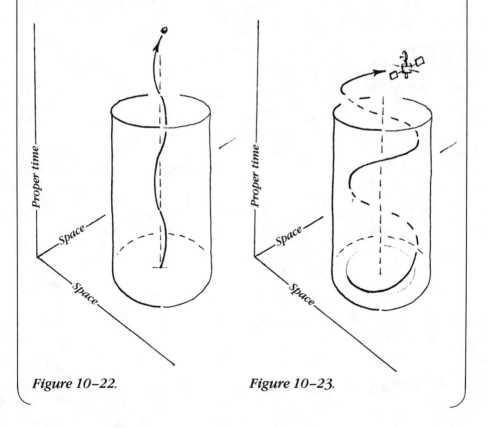

Figure 10–22.   *Figure 10–23.*

Figure 10-24 shows a light beam "refracted" by passing the earth, or better, the sun. Here the beam performs as though it passed through a cylindrical lens.

Figure 10-25 shows a blowup of a little cube of spacetime. If the goblet is simply dropped downward it follows path *a*, but if it is thrown sideways it follows path *b*. The curvature of the goblet's track through spacetime depends on the gradient of the "index of refraction" in the downward direction. The goblet on path *a* and the goblet on path *b* start moving through spacetime perpendicular to this gradient. Thus the radius of curvature of path *a* and path *b* are the same. It matters not whether you drop a goblet or throw it sideways, the radius of curvature of its path in spacetime, BUT CERTAINLY NOT IN SPACE, is the same!

Also note that goblet *a* requires more proper time than goblet *b* to fall a given distance. But both require the same amount of coordinate time, since the path lengths along paths *a* and *b* are equal.

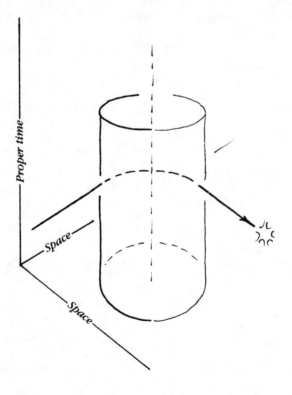

Figure 10-24.

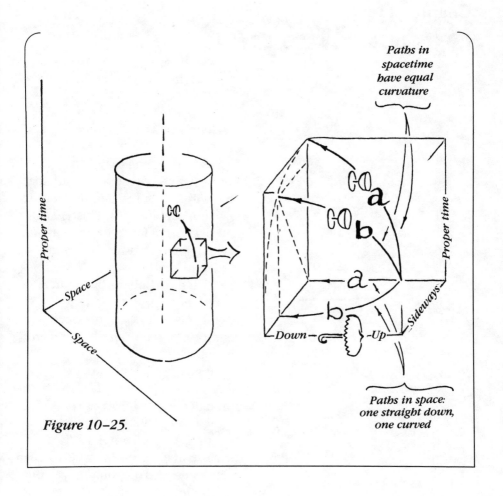

Paths in
spacetime
have equal
curvature

Proper time

Proper time

Space

Space

Sideways

Down — Up

Paths in space:
one straight down,
one curved

Figure 10–25.

Chapter 11
Space Warped

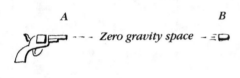

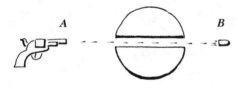

Figure 11–1.

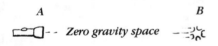

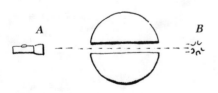

Figure 11–2.

Bullet Versus Photon

Now I am going to tell you something that is not obvious and that you will find surprising. Indeed, it was not obvious to Einstein. He had been thinking about gravity for several years before he came to realize it.

If you fire a bullet from point *A* to point *B* in empty zero-gravity space, it will travel at constant speed and require a certain time to complete the journey.

Suppose you put the earth between the points, drill a hole through it so the bullet can pass, and shoot again. The second bullet takes less time to go from *A* to *B*. Why? Because it falls on its way to the earth's center. True, on the way out the bullet loses all the speed it gained on the way in, but it was going faster inside the earth than it would have been going in empty space.

Now repeat the exercise with a flashlight rather than a revolver. The outcome is reversed. When the

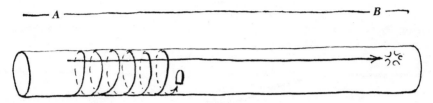

Figure 11–3. This sketch depicts the spacetime path of the bullet and the photon passing between points A and B in empty, zero-gravity space. The path length of the photon, which travels purely through space, determines the space between A and B.

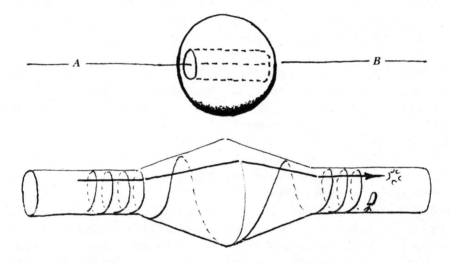

Figure 11–4. This sketch depicts the spacetime path of the bullet and the photon passing between points A and B when the earth is put in between the points. The path of the photon, which travels purely through space, is increased by the flare. So the space distance between A and B must somehow be increased by passing through the earth.

flashlight is fired through the hole in the earth, the time it takes to get from *A* to *B* is not decreased. It is increased. Why is the time the bullet requires to pass through the earth reduced, while the time the light requires is increased?

A crude caricature of the situation is sketched in Figures 11–3 and 11–4. In the no-earth, zero-gravity situation, the spacetime surface is a cylinder. The bullet goes slowly (compared to light), so it moves mostly through time. It spirals along, drifting slowly to the right. But the light does not age. It goes straight down the cylinder, traveling only through space.

Now when the spacetime surface is flared out by gravity, the pitch of

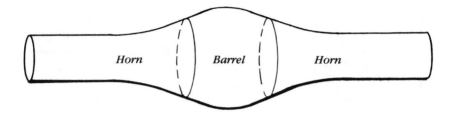

Figure 11-5. The spacetime surface in and near the earth, which is represented by two truncated cones in Figure 11-4, could more accurately be represented by a barrel with tapered horns extending from each end. The surface of the barrel is spherical and represents the part of the spacetime surface inside the earth. However, the cones are easier to make and to visualize, and produce the same gross effect as the barrel and horns.

the bullet's spiral spacetime path is increased. The pitch is increased, as previously shown, by riding down the conical section. Again, I suggest that you make a paper model and test it. Because of the increased pitch, the bullet wastes less of its speed spiraling through time. Its speed is better used to move it (right) through space. The total length of the bullet's track from A to B is decreased, which means its journey from A to B takes less time.

On the other hand, the path of light, which was previously straight down the cylinder, is now diverted over the conical flare. That makes the path and the time required to travel it longer.

Note that the travel time added for the light to go over the bump is finite. The travel time reduction enjoyed by the bullet might be infinite. If the bullet starts with zero velocity to the right, in zero-gravity space it will require infinite time to get from A to B. It will never get there. But if it starts with zero velocity and the

earth is between A and B, it will fall through in a finite time, thereby cutting an infinite amount from the travel time.

The Bump

Taken literally, the flare implies that the earth's gravity has somehow increased the distance between A and B. A string of field inspectors, called local observers, will find the number of miles or arm lengths between A and B to be increased when the earth is put between A and B.

How do the local observers perceive the extra space to be folded in between A and B? They don't perceive how it is folded. All they know is that it is there. The fold might be an up bump or a down bump or a bunch of wrinkles (Figure 11-8). Remember, only the internal fashion of the surface counts; its shape in space is just a display gimmick.

What about just stretching the wrinkles out by pulling A and B

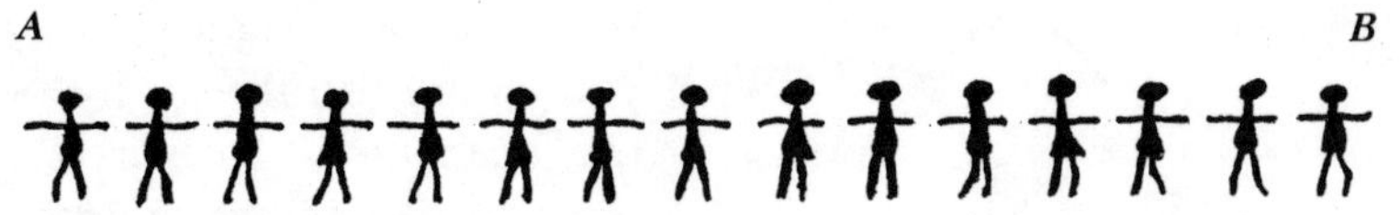

A   B

Figure 11–6. *In zero-gravity space, fifteen field inspectors reach from A to B.*

A   B

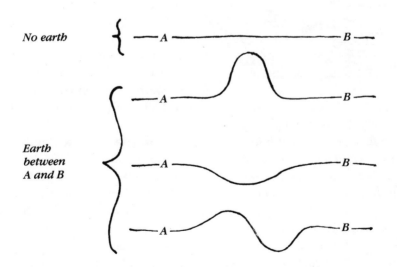

Figure 11–7. *With gravity, sixteen are required to reach from A to B.*

No earth

Earth between A and B

Figure 11–8.

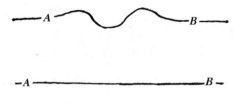

Figure 11–9.

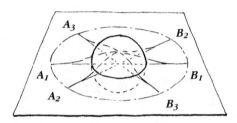

Figure 11–10. *The distance between all pairs of points* A *and* B *is increased when the earth is put between the points.*

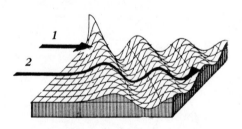

Figure 11–11. *You see the wrinkles in a surface when a light beam, which flies in a straight line, crashes into the wrinkles (Beam 1). If the beam flies in the surface, rather than in a straight line, there can be no crash (Beam 2). So you cannot see the wrinkles. Though you cannot see the wrinkles, they produce other effects that can be detected and will be described.*

farther apart? (Figure 11–9). Sorry, you cannot do that. Why? Suppose the distance is measured through the earth's center along various diameters of a great circle drawn around the earth. (You've got to have a drill in your imagination.) Each diameter will be slightly lengthened by gravity (Figure 11–10). But what about the circle's circumference? It is outside the region of gravity. (If you insist on being very precise, it is in a region of weak gravity.) So the circle's circumference is not affected. If you stretch *A* and *B* apart, you must also stretch the circle's circumference. But the circumference is unaffected by gravity, so it has no extra length folded into it. It ain't got no stretch. Hence, you can't pull out the wrinkles in the diameter. They are in there and you are stuck with them.

The neatest way you can arrange the wrinkles is to gather them into one bump or warp. But remember, the only reason for being neat is to make it easy for you to "see" what is going on. And also remember that you cannot really see the wrinkles or warp—though you might be able to feel them, like a blind ant feels a surface.

Normally you see wrinkles because light runs straight through space and hits the wrinkles (Figure 11–11). But you can't see wrinkles in this surface because the light is running in the surface!

And indeed it might be said that there really is no wrinkle or warp in the surface. For if there were a warp the surface could no longer pass through the center of the earth. But the surface does pass through the center of the earth. The surface just acts as if there were a warp in it. Essentially, the situation is not weirder than you imagine, it is weirder than you can imagine.

So gravity makes distance measurements come out **as if** there were a warp in the middle of the circle drawn around the earth. A lot of popular books on this subject show the warp as a down bump, caused by the earth's weight, and illustrate gravity as objects rolling down into the hole, as if there were an up and down in empty space! This is a powerfully misleading notion, so strike it from your mind. Remember, there would be gravity due to the time-slowing effect alone, even if the space warp did not exist. In fact, Einstein thought of gravity as only a time-slowing effect for several years. The bump is an extra effect, which occurred to him during World War I while he was employed by the Kaiser Wilhelm Institute, a "think tank" operated by the German Reich (federal) government.

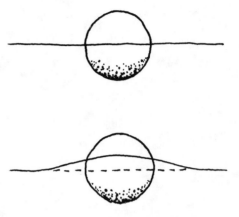

Figure 11–12. Even though distance measurements are made straight through the center of the earth, they measure **as if** *the measuring tape had a kink in it.*

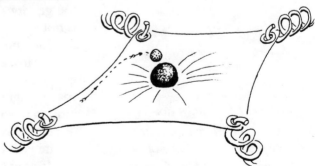

Figure 11–13. The wrong idea.

Another Cause

At first, slow time was thought to be just a side effect of gravity, but then it was realized that slow time could be thought of as the cause of gravity. So also was the warp in space first thought to be a side effect, but subsequently it was realized that the warp, by itself, could produce gravity, and that this gravity had to be added in to give the complete picture. How could the warp have been overlooked if it produces gravity? Because the warp and the extra gravity it produces are not easily detectable.

To detect any kind of wrinkle there must be sufficient warp in it. Shallow wrinkles in the finish of an automobile almost escape detection. The warp in space produced by the earth is presently virtually undetectable. To detect a wrinkle with only a slight curve in it, you need a vast expanse of space to work with. If you cut a small postage-stamp-sized patch out of an orange peel, it is practically flat. The warp only shows up when you handle a large piece of the peel. Which means that the gravity caused by curved space is not going to have a detectable effect on the motion of things in a room. Slow time rules there. But the gravitational effect of the warp in space might detectably affect the motion of something moving a long distance through strong gravity.

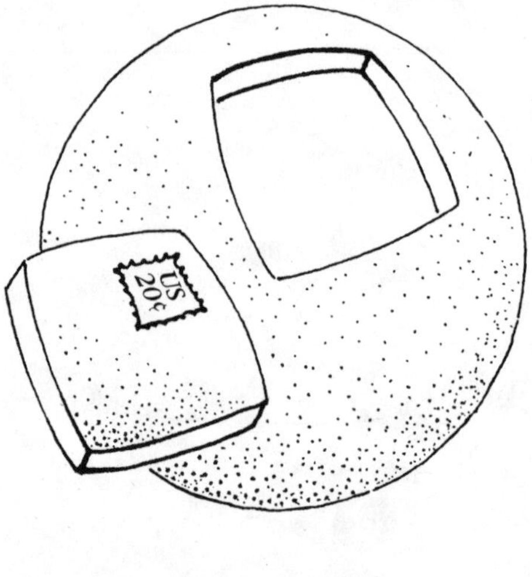

The effect you want to detect depends only on the warp, that is, the wrinkle in the path the thing follows

Figure 11–14. A small piece of orange peel can be laid approximately flat. A large piece of orange peel cannot be laid even approximately flat.

through space. The effect does not depend on how fast the thing traverses the path. However, the slow-time gravity effect does depend on how fast the thing moves. If it moves fast, slow time has less time to do its work. (That's why high-speed bullets go almost straight.) Therefore, if you want to detect the effect of the wrinkle, apart from the effect of slow time, it is best to choose something that moves through space very fast. Choose light.

The Bump's Effect

Now, disregard the slow-time aspect of gravity. That is, disregard the aspect of gravity you are familiar with in everyday life. Look at the warp effect alone.

Make a circle around the center of a large mass like the sun. Local observers will find that the surface in the circle measures as if it has a warp in it. I want you to make a model of this surface. Since it is hard to make a model of a nice smooth bump, suppose the bump is a cone.

There are two ways to make a cone. You can make it like a coffee filter. Simply make a circular paper disk. Cut it along one radius. Overlap the cut edges and tape them in place. Or you can make it as a tinsmith does. Make a circular disk and then cut a pie-shaped segment out of it. Then bring the two sides of the pie-shaped segment together and tape them. The disk will automatically buckle into a cone. Then stand the cone on a flat piece of paper and you have it.

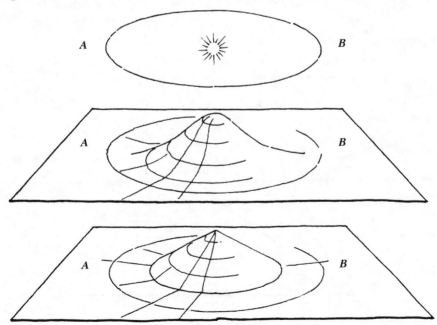

Figure 11–15.

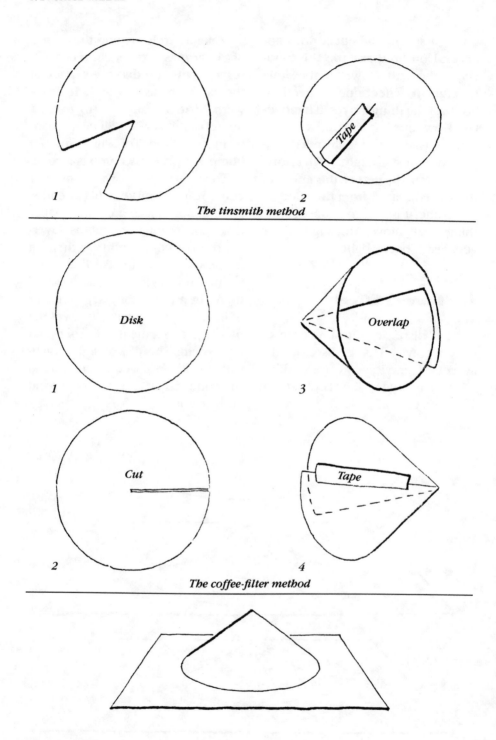

1

2

The tinsmith method

Disk

1

3

Overlap

Cut

2

4

Tape

The coffee-filter method

Figure 11–16.

Next, draw a line over the flat paper representing a beam of light that will pass close to, but not hit, the sun. The sun is at the apex of the cone. How should you extend the ray onto the conical surface? Just press the cone flat in the region of contact and extend the ray, in the direction it is going, a short distance on the cone. How should you draw the ray over the conical surface? Well, the ray goes straight but the cone is curved, so you have to reopen the cone, lay it flat, extend the short segment across the cone with a ruler, and then tape it all back together again. It is quite important to lay the cone flat while drawing the straight line and to draw with care, for you are, in effect, playing God. How should you extend the ray back onto the flat surface again? The same way you extended it onto the conical surface, by pressing the cone flat near the contact and extending the ray straight across. Lo and behold, the sun has deflected the light ray passing it.

The deflection you have just caused adds to the deflection due to the slow-time effect. It adds on enough to just double it. So the deflection of a light ray passing the sun is just double the amount Soldner and Einstein first thought it would be.

Note that the space warp will bend the path of something that is already traveling through space in order to make it travel in another direction through space. However, the space warp will not bend the

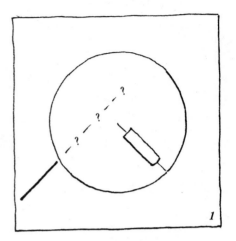

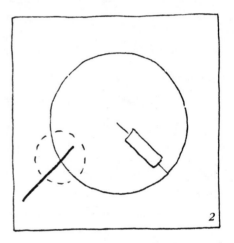

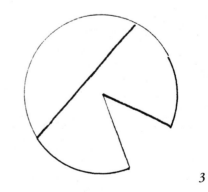

Figure 11–17, continued.

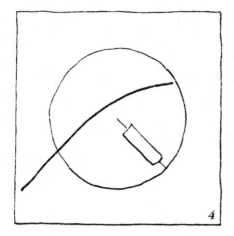

track of something that is moving only through time (stationary in space) in order to make it begin to move (down) through space.* When you drop a rock it is the slow-time aspect of gravity, not the space-warp aspect, that makes it fall.

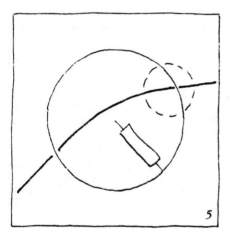

Figure 11–17, continued.

---

*This is reminiscent of the action of a magnetic field on a charged body. If the charge is not moving there is no force.

Question

The light ray was tracked over an up-bump space warp. If it were tracked over a down bump, its deflection would be:
a the same as it was for an up bump.
b opposite.

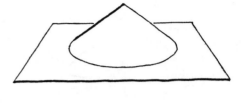
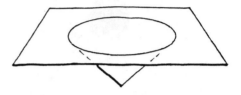

Figure 11–18.

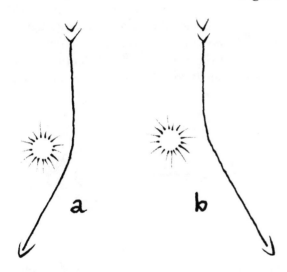

Figure 11–19.

Answer

The answer is **a.** If you doubt it, redraw the track on the inside rather than on the outside of the cone. Better yet, draw on transparent paper so that the same track can be seen from "top" and "bottom."

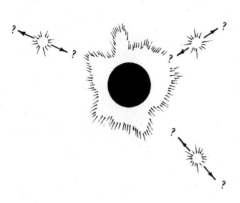

Question

During a solar eclipse you can see stars near the sun. The stars seem to be pulled:

a toward the sun.

b away from the sun.

Answer

The answer is **b.** The starlight is deflected by the sun's gravity, as illustrated. The rays are bent toward the sun, and so to you they seem to be coming from a position away from the sun! The deflection is too small to be detected by the naked eye.

Figure 11–20. The maximum solar deflection of passing starlight is about 2 seconds of arc. The very best resolution of a good naked eye is about 2 minutes of arc.

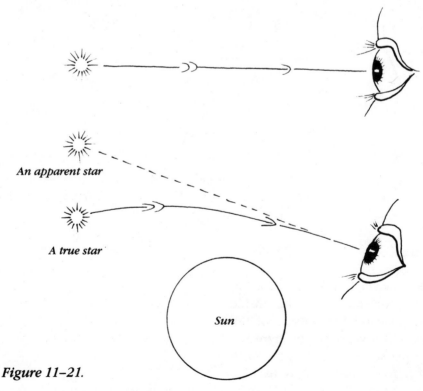

An apparent star

A true star

Sun

Figure 11–21.

Everything Wrinkled

If the space bump messes up the path of a light beam passing near the sun, should it not also mess up the path of other passing things? Indeed it does. It messes up the path of everything that goes over the wrinkle. In particular, it messes up the orbits of the planets. The effect is strongest on the planet Mercury, which is nearest the sun and therefore in the most warped part of the solar system. What is the effect?

The effect is exactly what you would expect. Due to the slow-time effect of conventional (Newtonian) gravity, the planet should trip around the sun in an elliptical path, and it does. But the space it runs in is not flat; it has a warp in it. Suppose, again, a conical bump centered on the sun. To make the flat paper into a cone, slit it to the sun's center and slip one cut edge slightly over the other (Figure 11–22). The ellipse can no longer close on itself. If the planet orbits the sun many times, the high point of the ellipse gradually shifts around in the same direction the planet is circling the sun.

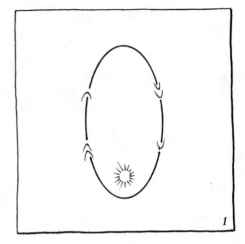

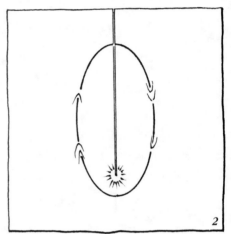

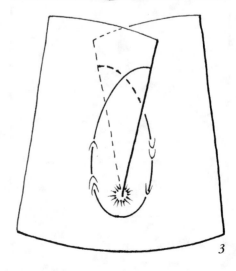

Figure 11–22.

Figure 11–23.

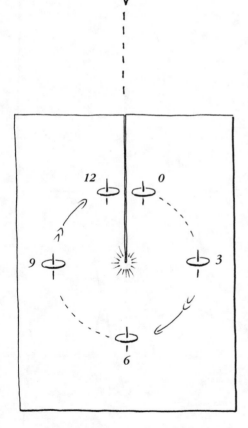

Figure 11–24.

The shifting of Mercury's orbit had been observed a century before Einstein was born, but went unexplained until the space wrinkle was recognized. Before Einstein, some astronomers supposed the shift was due to the gravitational effect of an undiscovered planet, Vulcan, in the space between Mercury and the sun. They spent a lot of effort looking for Vulcan, yet it was never found.

Could the wrinkle be detected if the orbit were perfectly circular? Yes.

How can you tell when a spacecraft goes once around a circular orbit? There are two ways to do it. Way I, by a local reference. Put a gyroscope in the spacecraft. If the gyro is free, it will always point in a fixed (local) direction. So if the gyro points at the sun now, it should point at the sun again after the spacecraft completes one orbit around the sun. Way II, by a distant reference. If the star Regulus is now opposite the sun, as viewed from the spacecraft, it should return to opposition again after the spacecraft completes one orbit around the sun.

Suppose you start off with both Regulus in opposition to the sun and the gyro pointing at the sun. After a while, Regulus returns to opposition and the gyro to pointing at the sun. But both don't recur together! Regulus returns before the gyro. How come?

Your expectations were based on flat space, but there is now a conical bump around the sun. Again make a

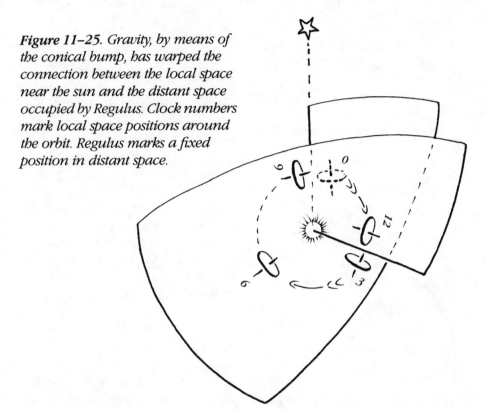

Figure 11–25. Gravity, by means of the conical bump, has warped the connection between the local space near the sun and the distant space occupied by Regulus. Clock numbers mark local space positions around the orbit. Regulus marks a fixed position in distant space.

paper model of the warped space by slitting the flat paper and slightly overlapping the slit edges. Behold! When the space is warped, returning to Regulus does NOT return the gyro to its original pointing. The spacecraft must move farther along its orbit before the gyro again points to the sun. That "little more" is the "mark" of warped space.

Question

Suppose the gyro was not pointed at the sun. Suppose it was pointed perpendicular (at right angles, 90°) to the plane of its circular orbit around the sun. On completing one orbit around the sun (returning to

Regulus), it would still be pointing in the same direction.

a True.

b False.

Answer

The answer is **a**. Slit open the conical warp and lay it flat. This surface is the space in which the spacecraft lives. In this space the gyro points in a fixed direction. That direction is perpendicular to the surface. The gyro is perpendicular to this surface, not only when the spacecraft carries it back to its starting point, but also throughout its orbit. This surface is also the plane of the spacecraft's orbit.

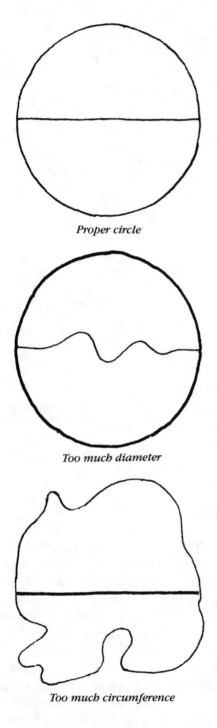

Proper circle

Too much diameter

Too much circumference

Figure 11–26.

Question

The effect of putting a mass inside a circle is to make the diameter of that circle measure longer than it would without the mass. But pretend that the effect of the mass is to make the diameter shorter than it would be without the mass. Then, in that imaginary case, the high point of a planet's orbit would:

a gradually shift around in the direction the planet was circling the sun.

b gradually shift around in the opposite direction.

Answer

The answer is **b.** In the real-life case, there is too much diameter for the circumference or too little circumference for the diameter. So you have to cut out or overlap out a pie-shaped segment of the orbital plane. But in the case supposed here, there is too little diameter for the circumference or too much circumference for the diameter. So you have to add in, rather than cut out, a segment. Figure 11–27 shows that its effect on the planet's orbit will be opposite to the effect of cutting out a segment. Its effect on the gyro and on the deflection of starlight passing the sun will also be opposite. But, of course, this is not the kind of warp found in nature.

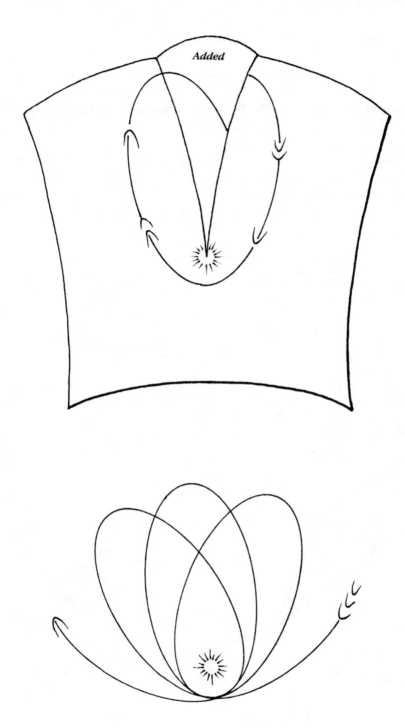

Figure 11–27.

How to draw straight lines on curved surfaces. How do you draw a straight line on a curved surface? If you can do it you can experiment with all kinds of ideas.

Any curved surface can be made up from or approximated by a number of flat surfaces put together. You can certainly draw a straight line on a flat surface. The trick is to draw a straight line around a bend in the surface. What do you do when the line comes to a bend?

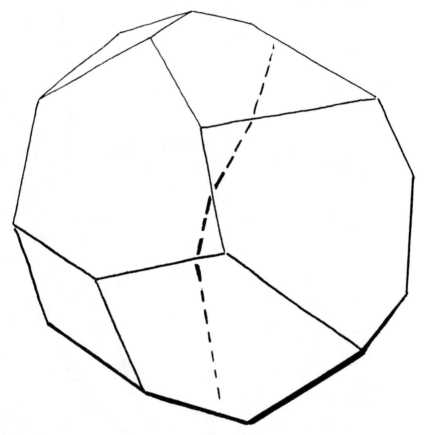

Figure 11–28.

You put a patch of paper around the corner. Extend the line on the paper patch. Remove the patch and flatten it. Extend the line over the flattened patch. Replace the patch on the corner. Extend the line over the adjacent flat surface. This is in effect like running a piece of tape around the corner of a box in such a way that it does not tear, buckle, or fold over on itself.

Now you have a nice new idea in your mind. What can you do with it?

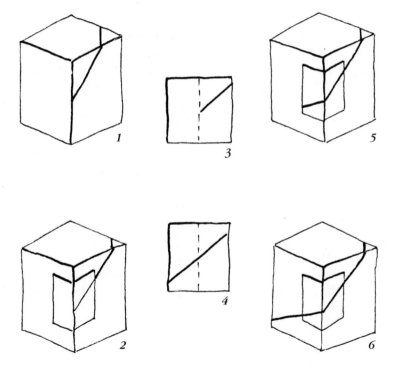

Figure 11–29.

Chapter 12

The Ends of Space and Time

Black Hole

Suppose the earth's mass could be increased at will. Time near the earth would go slower and slower. How slow could it go? In the extreme case it could stop. When time stops, nothing, not even light, moves. No motion means no escape, a perfect trap for everything, even light, a black hole.

But how can you make time stop? By flaring out the spacetime surface. In Figure 12–1, the circumference of the surface at B is double the circumference at A, so the speed of proper time at B is half the speed at A. At C the circumference is five times the circumference at A, so the

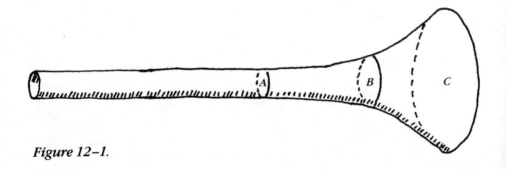

Figure 12–1.

Black hole anticipated. You should not be overly surprised that in the extreme case slow time can stop light. By extrapolation you can see that there might be some point S at the center of curvature of the falling light beam where the wave front would not move! Point S is not the center of the black hole. It is just a place near enough to have frozen time.

The gravity bending the illustrated beam is extremely strong, but not infinitely strong (were it infinitely strong the beam would fall straight down). Since the gravity is not infinitely strong the mass that makes the gravity need not be infinitely large. Neither infinite mass (nor infinite density) is required to make a black hole.

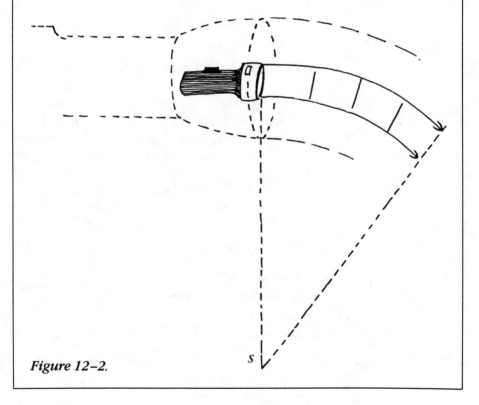

Figure 12–2.

speed of proper time at *C* is one-fifth the speed at *A*. To stop proper time, just let the spacetime surface flare out to infinity.

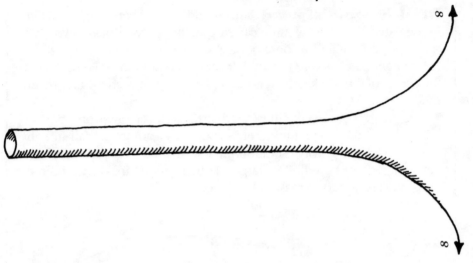

Figure 12–3.

As the mass of an object increases, the spacetime and spacespace surfaces evolve, as illustrated in Figures 12–4 to 12–7. In the limiting case, Figure 12–7, the diagrams rupture open. That means that an object that falls into a hole bored through such an extreme mass never comes out. It is indeed a black hole.

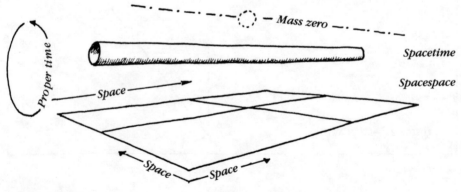

Figure 12–4.

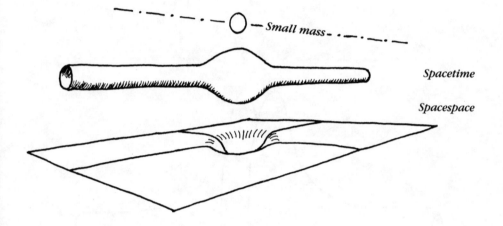

Figure 12–5.

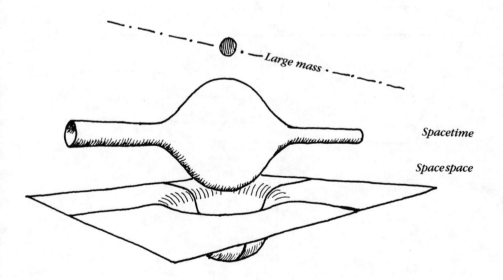

Figure 12–6.

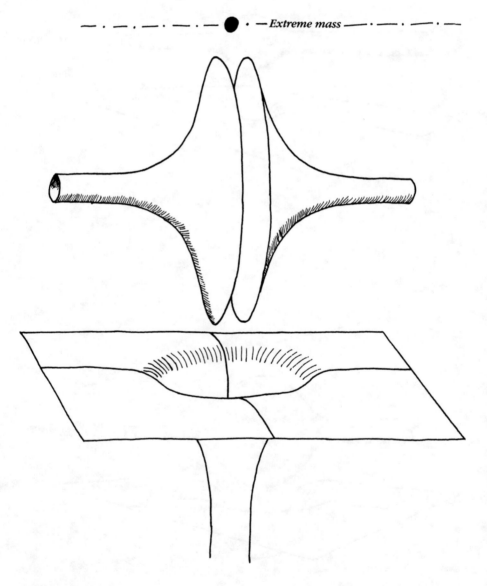

Figure 12–7.

Now I am going to astonish you. Were I to fall into the black hole, you would see me falling forever. That is, you would never see me reach the place, the horizon, where time stands still. But reckoning by my personal proper time, I would not fall forever. I would reach the horizon in a finite time. This dramatically illustrates the distinction between your time, coordinate time, and my personal proper time, or biological age.

Coordinate time, you will recall, is measured by the length of the track on the spacetime surface, while proper time is measured by the track's progress around the surface, that is, how many degrees it has wrapped around the surface. The sketch shows a track starting from rest at A and running to the horizon at infinity. The track is infinitely long, but it requires only a quarter (90°) turn around the spacetime surface to get to infinity!

What has been laid out here about black holes is what theory says such extreme masses will do to spacetime. However, the theory does not reveal whether or not such extreme masses truly exist in the world. Existence is a separate question. You can be most knowledgeably informed about that question by following the news notes in current issues of *Sky and Telescope* and *Scientific American.*

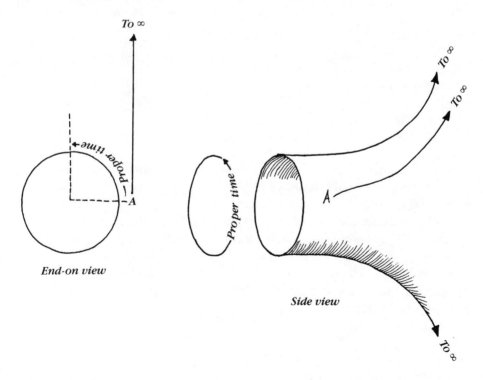

Figure 12–8.

How to draw infinity. How can you graphically represent infinity? If you draw a number line you always run off the page before you get to infinity.

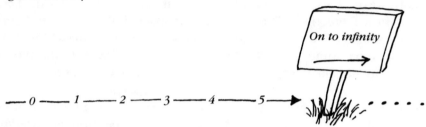

So do what artists do. Let some point on the page be infinity and draw the number line like this.

Figure 12–9.

How do artists get away with it? They just mimic a projection process that takes place on the retina of your eye. They call it perspective drawing and the sketch shows how it works.

The railroad track is the number line and it runs on to infinity. The artist projects the tracks onto the drawing in such a way that each point on the track corresponds to a point on the drawing. The points at infinity mark the horizon line of the drawing. The points on the part of the spacetime or spacespace diagram that are at infinity are likewise said to be on the horizon.

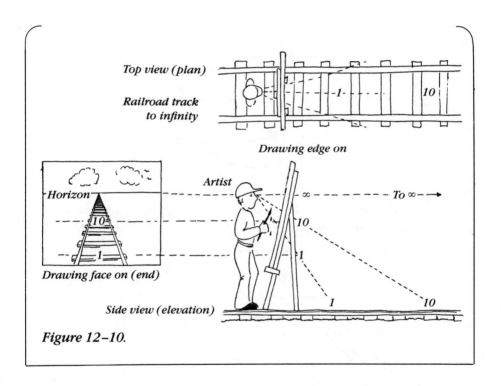

Top view (plan)

Railroad track
to infinity

Drawing edge on

Artist

Horizon

Drawing face on (end)

Side view (elevation)

Figure 12–10.

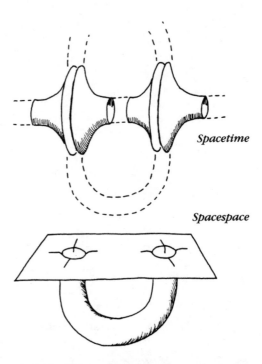

Spacetime

Spacespace

Figure 12–11. *A worm hole. When thinking about these things, you must ask not only if they can exist, but also if they can last. For example, a pencil can stand on its point, but can it remain standing on its point?*

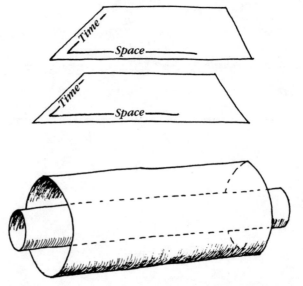

Figure 12–12. *Parallel universes, flat and rolled.*

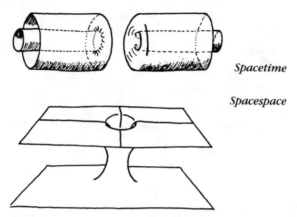

Spacetime

Spacespace

Figure 12–13. Bridge between parallel universes.

Worm Holes and Bridges

Even though the existence of black holes is uncertain, imagination is fast at work on the loose ends. True scientists, after all, are artists, and artists hate to have loose ends dangling. So all kinds of fanciful topological schemes are devised to tie in the spacetime diagram's loose ends. But be mindful that nothing in theory requires the loose ends to be tied in.

Here are two schemes for disposing of the loose ends. The first is called a worm hole. It joins two black holes together. You go down one and come up the other at a different place and at a different time.

The second scheme is called a bridge or gate to another universe. First you need two universes, that is, two separate worlds, each with its own space and time. These can be pictured as two parallel sheets of spacetime that never touch each other. The sheets can, if you wish, be rolled one inside the other without touching. (Don't forget that all this 3D stuff is just a display gimmick.) Finally, I can let the outside surface rupture inward and the inside surface rupture outward. This forms a bridge between the universes. You go down a black hole in this universe and emerge from a black hole in another universe.

Having eliminated the loose ends at the black hole, why not go on to eliminate the loose ends at the end of space and at the end of time? Again, all kinds of schemes can be conjured up. Some far-out notions involve one-sided surfaces called Klein bottles. I will show you two of the simplest schemes.

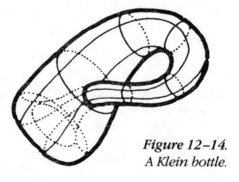

Figure 12–14. A Klein bottle.

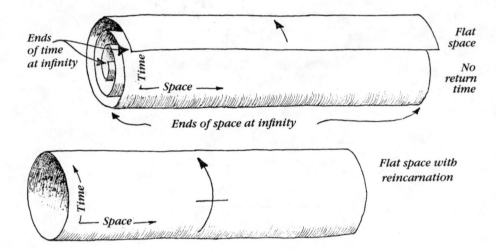

Ends of time at infinity

Time

Space →

Ends of space at infinity

Flat space

No return time

Flat space with reincarnation

Time →

Space →

Figure 12–15.

333

Figure 12–16. A mathematician thinks of a clock face as a circle of many loops, since 3 o'clock this morning, this afternoon, and tomorrow are all different times.

Two Cosmologies

The scroll of time may be presumed to have two loose ends, one infinitely far in the past and one infinitely far in the future. This may well be how it is. But if you don't like to run on forever, you can join the ends. Reincarnation! However, there is nothing in physics on which this joining can be based.

Space is also presumed to run on infinitely far in both directions. That Euclidean view may be correct. Yet if you don't like it, there are things you can do. You can join the ends forming a doughnut. Or you can tie the ends closed, forming a sausage that can be inflated into a sphere. These are closed universes. That is, you can sail around them like the Magellan Expedition sailed around the planet.

The spherical model implies that stationary objects, which track only through time, periodically crunch together and then expand away from each other. The get-together points can be pictured as the "north and south poles of spacetime."

Incidentally, you can join the ends of space without joining the ends of time, if you wish. If you don't join the ends of time, you have what mathematicians call a dough-nut of many sheets, or you can also make a sphere of many sheets.

Cosmology may be a game of topology. But do not forget that to play you must curve spacetime, and it takes mass to curve spacetime. Without the proper underlying amount and distribution of mass in the galaxies, you have nothing but a "might have been."

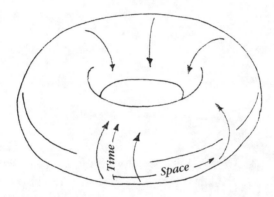

Figure 12–17. Bang-proof closed universe.

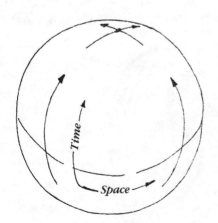

Figure 12–18. Oscillating universe.

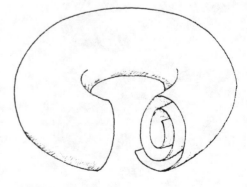

Figure 12–19. Universe of many sheets.

Question

There are two distinct possible kinds of doughnut universe. The distinction is based on the direction chosen for space and time. The two kinds of universe are sketched here.

a True.
b False.

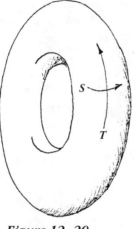

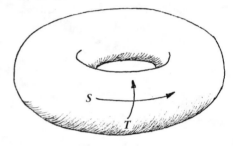

Figure 12–20.

Answer

The answer is **b.** The two kinds of doughnut are not distinct. One can be transformed into the other. How? Knit a doughnut with stripes running in one direction. Cut a little hole in it. Pull it inside out through the hole. You now have a doughnut with stripes running the other way.

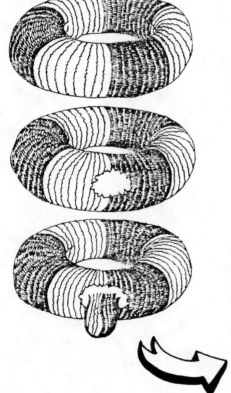

Figure 12–21.

TECHNICAL
APPENDICES

Peter Chung

Appendix A
Adding Speed

How to add speeds relativistically on the spacepropertime diagram

This appendix shows how to use the spacepropertime diagram to add speeds relativistically. The procedure is purely geometric. A cat is running on a moving yardstick. You know the cat's speed relative to the stick and the stick's speed relative to you. How do you find the cat's speed relative to you? This amounts to finding the sum of the cat's speed and the stick's speed. It's also equivalent to transforming from a frame where the yardstick is stationary to one where it's moving. When I say the stick is stationary I mean at rest in space. Though resting in space, it's moving through time.

To begin, suppose the stick is resting in space. This is represented in the first spacepropertime diagram. In the diagram the stick moves from L_0R_0 to L_1R_1 while the cat runs from C_0 to C_1. The stick has moved only through time. The cat has moved from the left to the right end of the stick, a distance $L_0R_0 = L_1R_1$ relative to you and also relative to the stick. Both you and the stick have aged an amount measured by the distance $L_0L_1 = R_0R_1 = C_0C_1$. The running cat has aged a lesser amount R_0C_1. The coordinate speed of the cat is L_0R_0/C_0C_1. The proper speed of the cat is L_0R_0/R_0C_1. Both speeds are relative to both you and the stick.

Now suppose the stick is set in motion to the right. This is most easily done by leaving the stick alone and setting you in motion to the left. In no way can your motion affect the amount of the stick's proper time or the cat's proper time required for the cat's run, and certainly your motion can't affect the cat's speed relative to the stick, though of course the cat's speed relative to you is changed. You want to find the cat's new speed relative to you. This is how you find it.

You draw a second spacepropertime diagram in which the stick and its spacetime path are tipped to represent its motion through space-

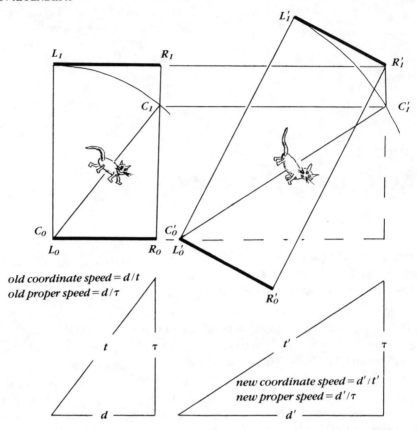

old coordinate speed = d / t
old proper speed = d / τ

new coordinate speed = d' / t'
new proper speed = d' / τ

Figure A–1. *The procedure for adding speeds, specifically adding the cat's speed to the stick's speed, is to transform the first diagram into the second diagram. In the first the stick is stationary, in the second it's moving. The cat's run begins at C_0 from the initial position of the stick's left end L_0. So these are the same point. Likewise for C_0' and L_0'. The cat's run ends at C_1 which must be someplace directly under R_1. Likewise, C_1' is directly under R_1'. You must bear in mind the following relationships. $L_0R_0 = L_1R_1 = L_0'R_0' = L_1'R_1'$ equals the stick's length. The vertical distance between the lines $L_0R_0L_0'$ and $L_1R_1R_1'$ equals C_0C_1. This distance represents the amount of the stick's proper time required for the run. The vertical distance between $L_0R_0L_0'$ and C_1C_1' represents the amount of the cat's proper time spent on the run. The distance $L_0'L_1' = R_0'R_1' = C_0'C_1'$ represents the amount of your time required for the cat's run when the stick is moving. When the stick isn't moving, there is no distinction between your time and the stick's time. Your time is called coordinate time. To aid visibility the lower triangles are abstracted from the upper diagrams. The cat's coordinate speed and proper speed both before and after the stick's speed is added are graphically deduced from the ratios between the triangles' sides.*

time. In the second diagram the length of the stick is unchanged, so $L_0'R_0'=L_0R_0$. The stick is tilted so its space dimension can be perpendicular to its path through spacetime. Desynchronization is a direct consequence of the tilt. When the cat departs from the left end of the stick, the stick is in position $L_0'R_0'$. When the cat arrives at the right end, the stick is in position $L_1'R_1'$. But that doesn't mean the cat is at R_1'. The cat hasn't aged as much as the stick. The cat is someplace directly below R_1' as previously C_1 was below R_1.

The cat departs from the left end of the stick at a certain proper stick age and arrives at the right end at a later proper stick age. The stick's motion can't alter the proper stick ages of these events. Therefore, L_0 and L_0' are on the same horizontal proper age line and R_1 and R_1' are on the same horizontal proper age line. This enables you to locate R_1' and draw $L_1'R_1'$ parallel and equal to $L_0'R_0'$. The distance $L_0'L_1'=R_0'R_1'$ represents how much of your time, coordinate time, was required for the cat to run the length of the moving stick. So the cat must be someplace on a circular arc of radius $L_0'L_1'$ centered on its starting point of $C_0'=L_0'$. The point on that arc directly below R_1' is the spacepropertime location of the cat when it arrives at the right end of the stick. Call that position C_1'. The line $C_0'C_1'$ gives the spacepropertime path of the cat. The path reveals the cat's speed relative to you.

This is not the only way to locate C_1'. The cat's proper age when it arrives at the stick's right end can't be altered by the stick's motion. Therefore, C_1 and C_1' must be on the same horizontal proper age line. The intersection of this line with the circular arc through L_1' or with the vertical line dropped from R_1' provide two more ways to locate C_1'.

To sum it all up, when the cat arrives at the right end of the stick you know three things: (i) the space location of the right end of the stick, (ii) the amount of your coordinate time required for the cat to run the length of the stick, and (iii) the amount of the cat's proper time spent running the length of the stick. These three things define (i) a vertical line through R_1', (ii) a circle through L_1' centered on L_0', and (iii) a horizontal line through C_1. Any pair of these things fixes the point C_1'.

It should be appreciated that even though the cat and the right end of the stick end up together at the same place at the same time, they are represented by different points on the diagram. One point is directly below the other. This is because they are at the same place in space. But the cat, having moved faster than the stick, aged less than the stick. This caused the vertical separation. What shows that the cat and the right end of the stick end up together is: one point is directly below the other, therefore, they have the same space location and the length of the cat's path on the diagram equals the length of the stick's path. Therefore, they arrived at the same time, the same coordinate time; coordinate time is your time.

If the amount of added speed is increased, or if the addition process is repeated, the cat's path on the diagram becomes ever more horizontal. But it can't get "more horizontal than completely horizontal," in which case the coordinate speed becomes the speed of light. So there is a limit to how much speed you can get by adding speeds. Of course that's one of relativity's central ideas.

Appendix B
Another View

How to transform from the spacepropertime to the spacecoordinatetime diagram

This appendix compares the spacetime diagram used in this book with the spacetime diagram most often used elsewhere.

The spacetime diagram in this book represents the speed of light as a horizontal line. The spacetime diagram in many other books represents the speed of light as a sloped 45-degree line. How come? Because the diagram in this book plots proper time against space. The diagram in the other books plots coordinate time against space. Which is right? Both are. They are different views of the same thing, like the plan drawing and elevation drawing of a house. Each has advantages and disadvantages. The important thing is to be able to refer back and forth from one to the other.

Before showing you how to refer back and forth, I first want to show you how to explicitly picture the relationship between space, proper time, and coordinate time. Suppose a spacepropertime diagram, the diagram used in this book, is drawn in the plane. See Figure B-1. Then, the path length O_pP_p equals your elapsed time, the coordinate time, for a traveler to move straight from O_p to P_p. (The little $_p$ means O_p and P_p are on the spacepropertime diagram.) This is because, according to the myth, the speed of all things through spacetime is constant; see Chapter 5.

Now imagine another dimension added to the diagram. And imagine the point P_p is projected in that dimension, perpendicular to the plane, a distance equal to O_pP_p. Then P_p will land on the surface of a 45-degree cone drawn through O_p. Call the landing point P. See Figure B-2. Now you can draw explicit space, proper time, and coordinate time axes. The space and proper time axes must be someplace in the plane of the spacepropertime diagram. The coordinate time axis is perpendicular to the plane, parallel to the cone's axis.

If you will suppose the traveler moves from O_p to P rather than P_p, then the projection of this motion on the spacepropertime diagram in the plane reveals the traveler's displacement in space and how much the traveler has aged during the trip, the traveler's elapsed proper time. The projection of O_pP on the coordinate time axis, perpendicular to the plane, equals P_pP, which measures how much of your time, coordinate time, has elapsed during the trip.

At P the traveler might change his speed through space and so change his direction of travel through spacetime. So P becomes a new O. And the traveler can take off in any direction on a new cone starting at P. If the traveler doesn't change his speed he just continues along the common tangent of the two cones. By going along a succession of many cones the traveler can go on all allowed paths through space, proper time, and coordinate time. See Figure B-3.

The cone also enables you to refer back and forth with ease between the spacepropertime diagram and the spacecoordinatetime diagram. See Figure B-4. The spacepropertime diagram used in this book is drawn in the space–proper time plane. On it I have drawn the "cosmic speedometer"—the quarter circle $Q_pP_pC_p$ and the needle O_pP_p, which represents the traveler's path on the spacepropertime diagram.

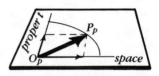

Figure B-1 is a spacepropertime diagram and O_pP_p is the needle of the cosmic speedometer, like the bold arrow in Figure 5-9. It can be resolved into a space and a proper-time component which tell how far the traveler has moved through space and how much the traveler has aged. The length O_pP_p measures coordinate time which tells how much you age as you watch.

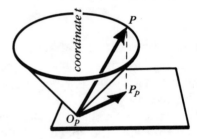

Figure B-2.

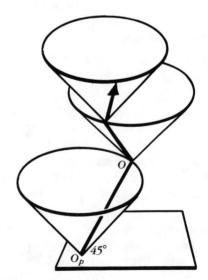

Figure B-3.

The spacecoordinatetime diagram used in other books is drawn in the space–coordinate time plane. On it I have drawn the 45-degree speed-of-light diagonal $O_c C_c$ and the hyperbolic arc $L_c R_c P_c$, which are important landmarks in the spacecoordinatetime diagram. The little $_c$ means these points are on the spacecoordinatetime diagram.

To transform from one diagram to the other draw the cone as illustrated. The angle $Q_c O_c C_c$ of the cone is 45 degrees. To move a point like R_p from the spacepropertime diagram to its corresponding point R_c in the spacecoordinate time diagram, draw a line from R_p perpendicular to the space–proper time plane. When the line touches the cone, drop it perpendicularly onto the space–coordinate time plane to locate R_c. Reverse this procedure if you want to go from R_c back to R_p. As R_p corresponds to R_c, so also Q_p, L_p, P_p, O_p, and C_p correspond to Q_c, L_c, P_c, O_c and C_c. Therefore, the horizontal speed-of-light line $O_p C_p$ becomes the 45-degree diagonal $O_c C_c$.

All points on the circular arc $Q_p P_p C_p$ have a coordinate time equal to the circle's radius $O_p Q_p$. The 45-degree cone ensures that these points transform to a line $Q_c P_c C_c$, the distance of which from O_c is $O_c Q_c$, which equals $O_p Q_p$. So, events which have the same coordinate time form a circle in the space-propertime diagram and a line in the spacecoordinatetime diagram. On the other hand, events which have the same proper time form a

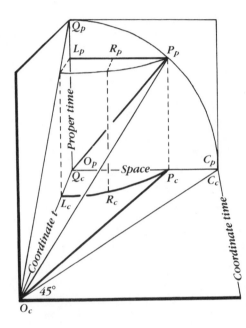

Figure B-4. The surface $O_c C_p P_p Q_p$ *is a quarter section of the cone. To aid visibility the cone is tipped so its axis is horizontal rather than vertical. And the spacepropertime diagram is projected from the plane passing through the cone's apex to the rear panel of the illustration. So* O_p *is moved from the cone's apex to the rear panel. The point in the corner formed by the three planes is called* O_p *when referred to the spacepropertime diagram, and called* Q_c *when referred to the spacecoordinatetime diagram. Likewise for* C_p *and* C_c.

line like $L_pR_pP_p$ in the spaceproper-time diagram and a hyperbolic arc $L_cR_cP_c$ in the spacecoordinatetime diagram. The arc is hyperbolic because the lines drawn from L_p and R_p form a plane which is parallel to the cone's axis, and a plane parallel to a cone's axis cuts the cone in a hyperbola, according to the original Greek definition of a hyperbola.

The spacecoordinatetime diagram is often called a Minkowski diagram. It was invented by Einstein's math teacher Minkowski about three years after Einstein came out with his ideas. The diagram is a complete graphic representation of Relativity. Incidentally, Minkowski didn't like Einstein as a student and Einstein didn't like Minkowski. He also didn't like Minkowski's diagram at first. But later he came to appreciate it, though he seldom used it.

Index

Wonderful Gift for a
TELESCOPE LOVER

Reproductions of fine old technical engravings from the age of brass and mahogany. Ideal for framing. Sizes range from one to two square feet. The small illustrations printed here cannot begin to do justice to the detail in the full-sized reproductions.

I enclose $_____ for the following order:

Print	A	B	C	D
Number of copies				

Name _____

School _____

Address _____

City _____

State _____ Zip _____

Astronomical Instruments Society
614 Vermont Street, San Francisco, CA 94107

1 print	$ 6
2 prints	11
3 prints	15
4 prints	18
5 or more prints	4 each

Tax and postage are included.
All orders must be prepaid.

A. The Medieval Astronomer
    11″ × 13″

B. The Great Telescope of the Lick Observatory
12″×18″

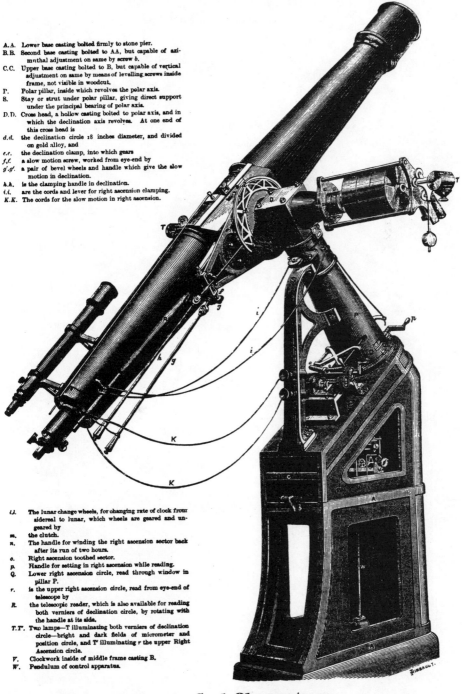

A.A. Lower base casting bolted firmly to stone pier.

B.B. Second base casting bolted to AA, but capable of azimuthal adjustment on same by screw *b*.

C.C. Upper base casting bolted to B, but capable of vertical adjustment on same by means of levelling screws inside frame, not visible in woodcut.

P. Polar pillar, inside which revolves the polar axis.

S. Stay or strut under polar pillar, giving direct support under the principal bearing of polar axis.

D.D. Cross head, a hollow casting bolted to polar axis, and in which the declination axis revolves. At one end of this cross head is

d.d. the declination circle 18 inches diameter, and divided on gold alloy, and

e.e. the declination clamp, into which gears

f.f. a slow motion screw, worked from eye-end by

g'.g'. a pair of bevel wheels and handle which give the slow motion in declination.

h.h. is the clamping handle in declination.

i.i. are the cords and lever for right ascension clamping.

K.K. The cords for the slow motion in right ascension.

l.l. The lunar change wheels, for changing rate of clock from sidereal to lunar, which wheels are geared and ungeared by

m. the clutch.

n. The handle for winding the right ascension sector back after its run of two hours.

o. Right ascension toothed sector.

p. Handle for setting in right ascension while reading.

Q. Lower right ascension circle, read through window in pillar P.

r. is the upper right ascension circle, read from eye-end of telescope by

R. the telescopic reader, which is also available for reading both verniers of declination circle, by rotating with the handle at its side.

T.T'. Two lamps—T illuminating both verniers of declination circle—bright and dark fields of micrometer and position circle, and T' illuminating *r* the upper Right Ascension circle.

V. Clockwork inside of middle frame casting B.

W. Pendulum of control apparatus.

C. The Equatorial of the Cork Observatory
13″ × 20″

D. 15-Inch Refractor of Mr. Wigglesworth's Observatory

13" × 20"

From time to time, the author of this book
Lewis Carroll Epstein
teaches short physics courses.

For class schedules contact the
University of California
Berkeley Extension
415-642-1061

MAGNET CAR

Will hanging a magnet in front of an iron car, as shown, make the car go?

a) Yes, it will go
b) It will move if there is no friction
c) It will not go

ANSWER: MAGNET CAR

The answer is: c. You could just dismiss the thing by saying that no work output will result from zero work input — or perpetual motion is impossible. Or you could invoke Newton's Third Law: the force on the car is equal and opposite to the force on the magnet — so they cancel out. But these formal explanations don't illustrate why it will not work.

To see intuitively why it will not work, improve the design by putting another magnet in front of the car. Then, to streamline things, put the magnets in the car. Then comes the question: which way will it go?

POOF AND FOOP

This is a stumper. If a can of compressed air is punctured and the escaping air blows to the right, the can will move to the left in a rocket-like fashion. Now consider a vacuum can that is punctured. The air blows in the left as it enters the can. After the vacuum is filled the can will

a) be moving to the left
b) be moving to the right
c) not be moving

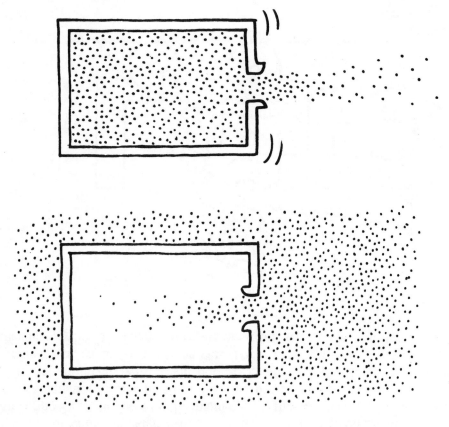

ANSWER: POOF AND FOOP

Are you reading this before you have formulated a reasoned answer in your thinking? If so, do you also exercise your body by watching others do push-ups? If the answers to both these questions is no, and if you have decided on answer c for POOF AND FOOP, you're to be congratulated!

To see why, consider the water-filled cart in Sketch I. We can see that it accelerates to the right because the force of water against its right wall is greater than its force of water against its left wall. The force against the left is less because the "force" that acts on the outlet is not exerted on the cart. Similarly with the can of compressed air. The "force" that acts on the hole is not exerted on the can, and the imbalance accelerates the can to the right.

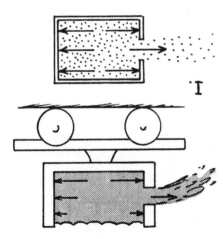

I.

Now consider Sketch II. Do these water-filled carts accelerate? No. Why not? Because the "force" of escaping water *is* exerted on the carts — on the outer wall of the top cart and on the inside wall of the bottom cart. So the water exerts no net force on the carts and no change in motion takes place (except for a momentarily slight oscillation about the center of mass). Likewise with the punctured vacuum can. The force of air that does not act at the hole nevertheless is exerted on another part of the can — on its left inner wall. So like the double-walled cart, the forces are balanced and no rocket propulsion occurs.

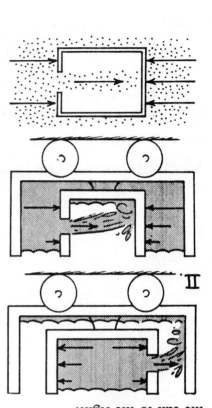

WORLD WAR II

To strike the tank factory the Flying Fortress should drop its bomb load

a) before it is over the target
b) when it is directly over the target
c) after passing over the target

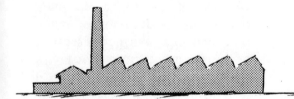

ANSWER: WORLD WAR II

The answer is: a. When the bomb is dropped it does not simply fall vertically to the ground straight below — such motion would require a zero horizontal velocity. When the bomb is released it has an initial horizontal component of velocity equal to that of the Flying Fortress (B29). If air resistance is negligible the dropping bomb continues forward with the airplane's velocity. The film shows how the bomb trails along directly under the aircraft. The bomb actually trails behind somewhat because air resistance is not really negligible. But if you drop a coin while inside the moving plane, then air resistance is negligible and the coin will fall at your feet. This is because its forward component of velocity is the same as that of your hand and your feet. It keeps up with the moving airplane. But if the plane accelerates while the coin is dropping, the coin does not land directly below its dropping position. Why?

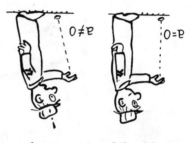

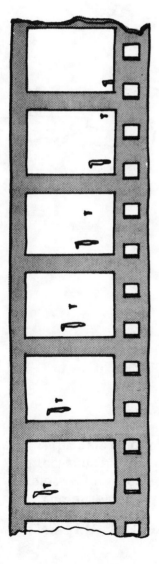

WHAT COLOR IS YOUR SHADOW?

On a clear and sunny day you are on snow and look at your shadow. You see that it is tinted

a) red
b) yellow
c) green
d) blue
e) not at all

ANSWER: WHAT COLOR IS YOUR SHADOW?

The answer is: d. The part of the snow in direct sun shows the color of the sun: yellow white. The snow in your shadow gets no direct sunlight, but is illuminated by light from the blue sky. Perhaps it is the blue color of shadows that makes people associate blue with cold.

E = MC²

The celebrated equation $E=mc^2$ or $m=E/c^2$ (c is the speed of light) tells us how much mass loss, m, must be suffered by a nuclear reactor in order to generate a given amount of energy, E. Which of the following statements is correct?

a) The same equation, $E=mc^2$ or $m=E/c^2$, also tells us how much mass loss, m, must be suffered by a flashlight battery when the flashlight puts out a given amount of energy, E.

b) The equation $E=mc^2$ applies to nuclear energy in a reactor, but not to chemical energy in a battery.

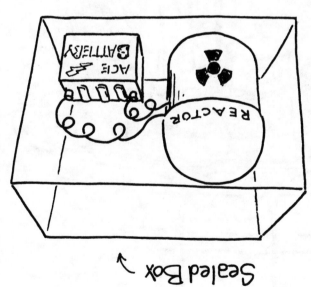

Sealed Box

ANSWER: $E=MC^2$

The answer is: a. If the mass-energy equation, $E=mc^2$, applies to any one form of energy, such as nuclear energy, then it must apply to every kind of energy, including battery energy. It is not hard to see why. Seal a nuclear reactor and a battery in a box. Nothing can enter or leave the box. Now let the reactor put out electric energy and let that energy be put into the battery. As the reactor puts out energy it must lose mass. But no mass can get out of the sealed box. So where could the reactor's lost mass be? The only other place it could be is in the battery. So the battery gains mass as it gains energy, and the battery loses mass as it puts out its energy. Whatever receives the battery's energy also receives some of the battery's mass.

BATTLESHIP FLOATING IN A BATHTUB

Can a battleship float in a bathtub?* Of course, you have to imagine a very big bathtub or a very small battleship. In either case, there is just a bit of water all around and under the ship. Specifically, suppose the ship weighs 100 tons (a very small ship) and the water in the tub weighs 100 pounds. Will it float or touch bottom?

a) It will float if there is enough water to go all around it
b) It will touch bottom because the ship's weight exceeds the water's weight

* This was my father's favorite physics question. —L. Epstein

ANSWER: BATTLESHIP FLOATING IN A BATHTUB

The answer is: a. There are a lot of ways to show why. This way was suggested by a student. Consider the ship floating in the ocean (sketch I). Next, surround the ship with a big plastic baggie — this is actually done sometimes with oil tankers — (sketch II). Next, let the ocean freeze except for the water in the baggie next to the ship (sketch III). Finally, get an ice sculptor to cut a bathtub out of the solid ice and you have it (sketch IV).

This question points out the danger of thinking in words, rather than thinking in pictures and ideas. If you just think in words you might reason: "To float, the battleship must displace its own weight in water. Its own weight is 100 tons, but there is only 100 pounds of water available — so it cannot float." But if you picture the idea you will see the displacement refers to the water that would fill the ship's hull if the inside of the ship's hull were filled to the water-line. And this displacement is 100 tons.

Don't rely on words, or equations, until you can picture the idea they represent.

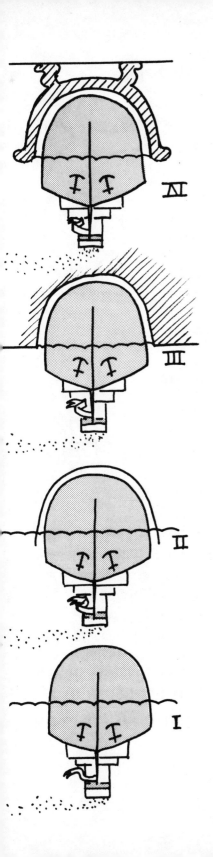

THIN AND FAT FILAMENTS

Light bulbs **A** and **B** are identical in all ways except that **B**'s filament is thicker than **A**'s. If screwed into 110-volt sockets,

a) **A** will be the brightest because it has the most resistance
b) **B** will be the brightest because it has the most resistance
c) **A** will be the brightest because it has the least resistance
d) **B** will be the brightest because it has the least resistance
e) Both will have the same brightness

THINKING PHYSICS is the TRIPLE POINT
where non-science, pre-med, and calculus physics
courses all meet. THINKING PHYSICS puts the
question where it logically—and historically—
belongs: before the explanation.